THE ART AND SCIENCE OF
WIND POWER

Wind Energy

David A. Rivkin, PhD
Laurel Silk, MEd, BA

JONES & BARTLETT
LEARNING

World Headquarters
Jones & Bartlett Learning
5 Wall Street
Burlington, MA 01803
978-443-5000
info@jblearning.com
www.jblearning.com

Jones & Bartlett Learning books and products are available through most bookstores and online booksellers. To contact Jones & Bartlett Learning directly, call 800-832-0034, fax 978-443-8000, or visit our website, www.jblearning.com.

Substantial discounts on bulk quantities of Jones & Bartlett Learning publications are available to corporations, professional associations, and other qualified organizations. For details and specific discount information, contact the special sales department at Jones & Bartlett Learning via the above contact information or send an email to specialsales@jblearning.com.

Production Credits
Chief Executive Officer: Ty Field
President: James Homer
SVP, Editor-in-Chief: Michael Johnson
SVP, Chief Technology Officer: Dean Fossella
SVP, Chief Marketing Officer: Alison M. Pendergast
VP, Manufacturing and Inventory Control: Therese Connell
SVP, Curriculum Solutions: Christopher Will
Director of Sales, Curriculum Solutions: Randi Roger
Editorial Management: High Stakes Writing, LLC, Editor and Publisher: Lawrence J. Goodrich
Managing Editor, HSW: Ruth Walker
Copy Editor, HSW: April Austin
Senior Editorial Assistant: Rainna Erikson
Production Manager: Susan Schultz
Production Editor: Keith Henry
Production Assistant: Kristen Rogers
Senior Marketing Manager: Andrea DeFronzo
Manufacturing and Inventory Control Supervisor: Amy Bacus
Composition: Cenveo Publisher Services
Cover Design: Kristin E. Parker
Rights & Photo Research Associate: Lian Bruno
Cover Image: © WDG Photo/ShutterStock, Inc.
Printing and Binding: Edwards Brothers Malloy
Cover Printing: Edwards Brothers Malloy

ISBN: 978-1-4496-2447-7

Library of Congress Cataloging-in-Publication Data
Unavailable at time of publication.

6048

Printed in the United States of America
16 15 14 13 12   10 9 8 7 6 5 4 3 2 1

Brief Contents

Contents

Preface

THE WIND ENERGY INDUSTRY is at the forefront of the world's shift away from reliance on fossil fuels. In just a few short decades wind energy has evolved dramatically. Technological advances now make wind energy a cost-effective solution for the world's ever-growing energy needs. The United States is now one of the world's leaders in overall wind power capacity.

As the wind energy industry continues to expand in the United States and around the globe, it will provide many opportunities for workers in search of new careers. These careers extend beyond the wind farm and include the efforts of employees who work in manufacturing plants, offices, and construction, as well as operation and maintenance. According to estimates from the American Wind Energy Association, approximately 85,000 Americans are currently employed in the wind-power industry. Despite the growing demand for skilled workers, there remains a lack of serious educational resources to meet the market's demand.

The *Art and Science of Wind Power* series was developed to fill this education gap. Each book in the series examines performance challenges using a systems perspective. Readers do not learn design and installation steps in a vacuum—instead they examine interrelationships and discover new ways to improve their own systems and positively contribute to the industry.

This series was developed for both the novice and expert. The texts take the learner from an overview of wind energy, through design and installation steps and considerations, to the design and installation of commercial wind systems.

Wind Energy provides readers with an overview of wind-energy history, basic concepts, costs, and uses. This book is designed for the novice and emphasizes an overview of wind-energy conversion and applications of various wind-turbine systems.

About the Authors

Prof. David A. Rivkin, PhD, is managing director and dean of the College of Science and Technology at the Sustainable Methods Institute (SMI), an online university and innovation center.

He is also the chairman of the Department of Nanosciences in Renewable Energy at Chiist University, Atlanta, Georgia; and dean of education and research at the Israel Sustainability Institute. He is also chief scientist and director at the Adamah Group and its wind power division, in Israel.

Professor Rivkin was the founder, associate professor, and chairman of the Green Technologies Department at Ohalo College of Katzrin; associate professor at National University, based in San Diego, Calif.; International Technological University, in San Jose, Calif.; and the Graduate School of Science and Technology at the Technical University of Munich, in Germany.

He holds dual bachelor's degrees in chemistry and nuclear engineering from the University of California at Berkeley. Professor Rivkin pursued postgraduate studies toward a doctorate in biophysics at the University of California and later completed a PhD in business sciences at the European School of Business London at Regent's College, with a focus on small business sustainability. He has over 25 years of professional experience in both industry and academia.

He has taught at internationally renowned colleges and universities in Europe, India, China, the United States, and Israel. In 2010 Professor Rivkin was nominated to be a Fulbright scholar. He is also an Institute of Electrical and Electronics Engineers (IEEE) senior member and distinguished lecturer, a principal adviser in clean technologies to the National Science Foundation, a certified program manager, a certified corporate sustainability expert, and a certified green energy professional.

He is the winner of numerous technical and managerial awards and has been recognized for his outstanding contributions by governments as well as the United Nations. A serial entrepreneur, with roles in several successful ventures, including as founder and chief of technology for SciEssence International, Professor Rivkin has a multidisciplinary background, from biosciences to nanotechnology from health to energy, that gives him broad expertise in sustainability.

Laurel Silk has managed e-learning initiatives at three leading universities in Arizona including Arizona State University, University of Phoenix, and Grand Canyon University. Among the highlights of her career, she created a virtual doctoral library for research students and designed and implemented web-based doctoral studies programs in health care and education administration. Ms. Silk is a former classroom instructor, in which capacity she designed and taught courses in freshman English, world literature, and critical thinking. As vice president of SilkWeb, she has created undergraduate and graduate online courses in higher education, renewable energy, business, and nursing. Ms. Silk holds a Masters degree in education with a focus on adult learning and instructional design technologies from University of Phoenix and has a Bachelor of Arts in English from Arizona State University.

History

chapter 1

THIS CHAPTER IS DEVOTED to giving you an understanding of the history of wind power. This review examines wind power and how it has been used in the past up until today. The primary uses of wind power and significant innovators of wind technology are also included. Take note that this chapter describes wind power's history in chronological order, ending with a quick look at what the future holds for wind energy.

Chapter Topics

This chapter covers the following topics and concepts:

- What types of early wind power existed
- How US windmills were used to pump water
- How wind generation of electricity has progressed from the late 19th century until today
- What trends are predicted for wind power technology

Chapter Goals

When you complete this chapter, you will be able to:

- Outline the major uses and developments of wind power up to the 18th century
- Explain how and why US windmills were adapted for various uses in the 19th century
- Outline how electricity has been generated by wind power starting in the 19th century
- Define the California wind rush and explain why it occurred
- Explain future trends in wind power technology

Early Wind Power

For many people, wind technology is a modern invention. Actually, early wind power has an extensive history, which some historians claim dates back 3,000 years. Authors have noted wind power's beginnings in the Mediterranean region, in China, the Middle East, and Europe. In the past, the primary uses for wind power included grinding grain, sailing ships, and pumping water.

The following is a chronological look at the early history of wind power:

- First century B.C.—The exact beginnings of early wind power are disputed. However, some experts claim that evidence of stone windmills exists in Egypt. They estimate that these windmills were built 3,000 years ago. Also noteworthy is the Greek engineer and inventor Hero or Heron of Alexandria, who some say developed the first windmill.
- Seventh century—Experts agree that windmills existed along the border of Persia and Afghanistan beginning in the year 644. These windmills, which rotated on a **vertical axis** FIGURE 1-1, were used for milling grain.
- Twelfth century—The first windmills built on a **horizontal axis** FIGURE 1-2 are documented in the province of Brabant (1119) and in the Duchy of Normandy (1180), both in northwestern Europe. These windmills were called **post windmills** and their use quickly extended throughout the northern and eastern regions of Europe, including even Russia. Post windmills consisted of a millhouse that revolves around a central post and a wind wheel with four sails. It was used to mill grain.
- Thirteenth and fourteenth centuries—Sometime during these two centuries, the **tower windmill** was developed (see FIGURE 1-3.) It differed from wooden post windmills in that the wind wheel was propped on a

FIGURE 1-1 A vertical axis windmill.
© iStockphoto/Thinkstock

round stone tower. Typically, the wind wheel consisted of triangular sails found in southwestern France and in the Mediterranean.

- **Fifteenth century**—The Dutch transformed the post windmill into the **hollow-post mill** FIGURE 1-4, which included a scoop wheel necessary for pumping water. The drive for the scoop wheel is found within the pyramid-shaped foundation. The primary uses of the hollow-post mill in

FIGURE 1-2 A horizontal axis windmill.
© iStockphoto/Thinkstock

FIGURE 1-3 Greek tower windmills of a type common in the 13th and 14th centuries.

FIGURE 1-4 A hollow-post mill popular in the Netherlands in the 15th century.

FIGURE 1-5 Typical Dutch windmills.

the Netherlands were for draining water, as well as for milling grain and sawing wood.

- **Sixteenth century**—Inventors in the Netherlands developed what is known as the **Dutch windmill** FIGURE 1-5. It functioned using a fixed base and a rotating tower cap that turned the wind wheel.
- **Seventeenth and eighteenth centuries**—Many developments occurred during these two centuries. Of the most significant was the introduction of scientific testing and evaluation of windmills. When inventors utilized mathematical and physical laws, many new and improved designs emerged. Windmills increased in efficiency when inventors Andrew Meikle and Edmund Lee created a fantail to produce automatic **yawing** of Dutch windmills. Meikle also invented windmills with **spring sails**, which allowed increased control over the speed and power of the windmill. Spring sails, initially made of wood, were shutters that hung on an iron rod.

NOTE

The Chinese also used vertical axis windmills to drain rice fields and pump water. It's not clear when they began using windmills, however.

Use of Windmills to Pump Water in the United States

By the beginning of the 19th century, the use of windmills was widespread in Europe. In the United States, known as the New World during this time, early settlers from the Netherlands and England constructed windmills along the

East Coast. But it wasn't until settlers moved west that innovations in windmill technology occurred.

The only water source in the extensive Plains region of today's US was found underground. The existing windmills, also known as fan mills, proved ineffective. Fan mills consisted of wind wheels constructed of less-than-durable sailcloth, which was often damaged in poor weather. This model demanded that owners monitor and untangle sails, especially during storms. But in the mid-19th century, Daniel Halladay, a mechanic from Connecticut, addressed this problem and created a windmill that pumped water efficiently without requiring as much attention as earlier mills did.

Self-Regulating Fan Mills

Halladay changed the design of the fan mill by replacing the cloth sail with many thin wooden blades, which completely covered the wind wheel. Additionally, he connected the wind wheel so that it hung on a ring instead of affixing it directly to the shaft of the windmill. He also incorporated flyweights, which helped control the movement of the ring. These changes allowed for the wind wheel to regulate itself in low and high winds. The combination of his changes resulted in a self-regulating fan mill that was ideal for pumping water. Halladay sold many of his new and improved wind turbines to the American railroad companies, which were used to keep water tanks replenished while in transit. This particular brand of windmill remained on the market until 1929.

A few years later, the Reverend Leonhard R. Wheeler of Wisconsin streamlined Halladay's original design. He incorporated a wind vane that operated in conjunction with a weight. This adjustment to the windmill rotated the wind wheel away from the wind when wind speed decreased. This would cause the wheel to return to its original placement. Wheeler's windmill, which he called "the Eclipse," established the model of the American wind turbine.

Wind Generation of Electricity

Prior to the 19th century, power generated by windmills was utilized for basic survival, in tasks such as milling grain and pumping water. In the late 1800s, the use of electricity began to grow rapidly. By the early 1900s most major cities in industrialized nations utilized electricity. Despite the growth of power plant technology, windmill use in Europe and wind turbines in the US maintained popularity. But it wasn't until the late 19th century that the first attempts at generating electricity via wind power occurred.

1888: Charles Brush

Charles Brush of the United States was the first inventor to use windmills to generate electricity, otherwise known as wind turbines. He is best known for creating the 12 kW DC windmill generator.

1891 to 1918: Poul La Cour

Shortly after Brush's innovation, Poul La Cour, in Denmark, made great advances in wind-generated electricity. At the time, rural areas largely lacked the same access to electricity as the large cities. With the government's support, La Cour found a way to have wind power fill the need. He based his wind turbine **FIGURE 1-6** on traditional windmills but opted to create different sails for the wind wheel. Traditional wind wheels had aerodynamically shaped sails. Instead, La Cour used a rotor made up with four shutter sails. His design was a success.

In 1897, La Cour went on to test, evaluate, and improve his designs in a wind tunnel and in a separate test station he set up himself. At the turn of the 20th century, the Lykkegard Company applied his designs, and by 1908, 72 of La Cour's wind turbines were in use in rural Denmark. Ten years later, that number rose to 120, partly due to the increasing cost of fuel during World War I. La-Cour-Lykkegard turbines came in different sizes and varying power outputs from 10 to 35 kW. After the war's end in 1918, the demand for wind power technology plummeted as fuel became affordable again.

FIGURE 1-6 A La Cour-designed wind turbine used to generate electricity.

The 1930s: The US Rural Electrification Administration

In 1930 in the United States, wind turbine manufacturing reached a high never seen before. More than six million wind turbines had been produced. But with the introduction of the Rural Electrification Program, which brought electricity rural inhabitants, the popularity of wind turbines declined quickly.

The 1940s: The Smith-Putnam Machine

The next decade saw US engineer Palmer Cosslett Putnam combine both large wind turbines and power plants to meet the public's utility needs. With the support of the S. Morgan Smith Company, a water turbine manufacturer, Putnam designed the first of what was considered a large wind turbine at the time **FIGURE 1-7**. It was constructed in Vermont, on a hill called Grandpa's Knob.

The Smith-Putnam wind turbine provided electricity for the Central Vermont Public Service Company for approximately four years. On March 26, 1945, a broken rotor blade stopped operations. Due to lack of funds, production was permanently shut down.

Putnam also provided the world of wind technology with data that still applies today. In 1947, he outlined the technical data for an ideally sized and

FIGURE 1-7 The original Smith-Putnam wind turbine in Vermont, 1941.
Courtesy of NASA

economical wind turbine in his book *Power from the Wind.* The data is as follows:

- **Rotor diameter**—53.3-68.5 m (175-225 feet)
- **Tower height**—45.7-53.3 m (150-175 feet)
- **Generator power**—1500-2500 kW

> **NOTE**
>
> The data associated with the Smith-Putnam wind turbine, the largest of its kind at the time, is as follows:
> - **Rotor diameter**–53.3 m
> - **Tower height**–35.6 m
> - **Generator power**–1250 kW

The 1950s–1960s: Ulrich Hütter

In Germany, Ulrich Hütter created a specific design of wind turbines that continues to impact wind turbine technology today. Initially, he created a small wind turbine that had a rotor diameter of 10 m and an output of 8 to 10 kW of rated power. It experienced moderate success. However, in 1958, he went on to design a much larger wind turbine, known as the W-34 (see **FIGURE 1-8**), named for the 34 m diameter of the rotor. Its rated power was 100 kW.

Other notable features of Hütter's W-34 that affect the design of today's wind turbines include:

- **Rotor blade material**—The blades were composed of a fiberglass material that went on to be the standard for producing gliders.
- **Teetering hub**—The structure that joined the rotor blades and the rotor and allowed for teetering movement, which was a more efficient and simplified mechanism than Putnam had created earlier.
- **Generator power**—It produced 100 kW, which was effective at maximizing the low wind speeds found in central Germany.

Despite its high-quality design, it operated for only 10 years and not at full capacity. Lack of funding forced the turbine to be dismantled in 1968.

The 1970s: The Reemergence of Wind Power Technology

In 1973, Western industrialized nations went into shock when crude oil prices multiplied within months. These countries sounded the alarms and announced an energy crisis. It is important to note that the energy crisis was not the result of lack of resources but solely due to rising prices demanded by the exporting countries. Other types of energy, in particular **renewable energy sources**, were sought out to lessen dependency on crude oil. North America and Europe turned to wind power technology. The countries that made the most significant developments in wind power technology included:

- **The United States**—In 1973, $200 million was allocated to the US Federal Wind Energy Program, which funded extensive studies and experiments that advanced wind power technology. The private sector also invested in the construction of advanced power plants.
- **Denmark**—In 1974, this country aimed to fulfill 10 percent of its power needs from wind energy. A large, out-of-use 200 kW wind turbine was updated, two large turbines were constructed, and private use of small wind turbines was

encouraged. As an incentive to incorporate wind energy, the government provided generous subsidies to businesses that purchased small wind turbines.

- **Sweden**—In 1975, the Swedish National Board for Energy Source Development (NE) was established; 280 million Swedish crowns (more than US $44 million) went into research and development of wind energy technology, as well as toward the creation of two large wind turbines, at 2 MW- and 3 MW-rated power.
- **Germany**—In 1974, the government began subsidizing research and experimentation of wind technology. One of the first projects undertaken

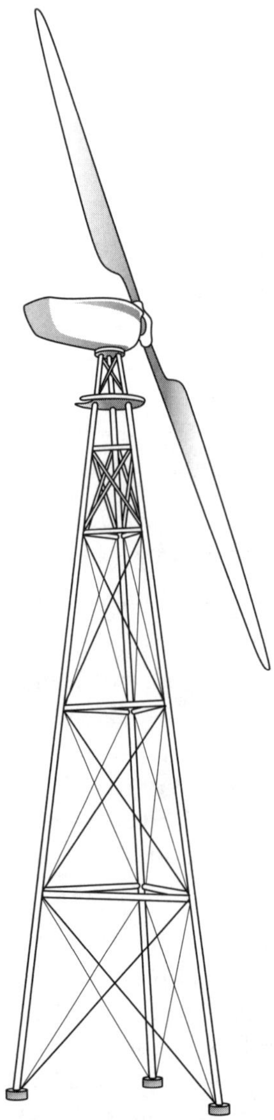

FIGURE 1-8 Hütter's W-34 wind turbine.

involved building a 3 MW wind turbine, but because of various external factors, such as politicians looking for fast fame, it was not successful. However, in subsequent years small wind turbine production prospered.

The California Wind Rush

In 1976, the California legislature provided private investors with a 10 percent tax credit to cover the expenses of using solar energy. In 1978, the federal government passed the National Energy Act, which included the same tax credit for private investors nationwide. Further legislation supported the use of renewable energy sources. The Public Utilities Regulatory Act (PURPA) required public utilities companies to make use of energy produced from renewable energy sources within their grid and to pay for a portion of it. In the mid-1980s in California, the Pacific Gas and Electric Company (PG&E) and the Southern California Edison Company (SCE) covered costs up to 10 percent of the power channeled into their grids.

> **NOTE**
>
> The Canadian federal government also encouraged the exploration of wind power technology as a viable energy source.

The first wind farms in California appeared in 1979 but due to poor planning and execution they didn't gain in popularity until 1981. In the early 1980s, wind turbines were installed daily on new farms and Danish manufacturers exported many wind turbines. The number of wind turbines in California continued to grow until 1985. By this time, 40 percent of wind turbines were of Danish origin.

In the late 1980s, the combination of dropping oil prices and cheaper costs to run power plants led the U.S. government to stop providing tax credits. Taken together, these factors made wind farm technology much less attractive. Wind turbine manufacturers continued to build and sell their products. Fortunately, they were able to offer them at lower prices, which kept some wind farms in operation.

> **NOTE**
>
> The sudden interest and economic potential of wind farms in the 1970s in California is sometimes referred to as the California wind rush because of the similarity it shares with the California gold rush of the 19th century. When word got out in the mid-1800s that gold could be found in California, thousands of people flocked there hoping to strike it rich. The fervor in which wind farms were constructed in California in the 1970s shared the same excitement and hope for good fortune.

FUTURE TRENDS IN WIND POWER TECHNOLOGY

Are wind farms in our future? Definitely. One of the latest trends in wind power technology is offshore wind farms **FIGURE 1-9**. As of 2010, Europe had more than 900 wind turbines operating on 43 offshore wind farms. China's first offshore wind farm began operation in 2010. The United States has yet to implement one. Prior to 2009, the US government restricted the construction of offshore wind farms up to 3.5 miles (5.6 kilometers) from the shoreline.

(Continues)

FUTURE TRENDS IN WIND POWER TECHNOLOGY (Continued)

Wind developers celebrated a new federal law in 2009 that allows wind farms to be built up to 230 miles away from the shore. Wind power experts say that more energy can be generated when wind turbines are far from land.

FIGURE 1-9 An offshore wind farm.
© T.W. van Urk/ShutterStock, Inc.

In April 2010, the US government authorized the construction of the country's first wind farm. It is to include 130 wind turbines and be built off Cape Cod, on the Massachusetts coast.

Another trend worth mentioning is that when it comes to wind power, bigger isn't always better. As costs to produce wind turbines increase, mostly from the high prices of steel and other raw materials, wind power specialists recognize the turbine's size doesn't need to increase. Instead, by altering materials and structure, more energy can be generated from smaller machines. Wind power manufacturers at General Electric (GE) are currently exploring how to do exactly that. Meanwhile, smaller wind turbines designed for factories, office buildings, and even homes are attracting interest. Some countries are proposing laws that would require 1 kW wind turbines be installed on all available rooftops in new residential construction, driving a new distributed wind power paradigm similar to that currently in place for solar panels, both for electricity and for hot water.

CHAPTER SUMMARY

A look at the history of wind energy reveals that wind power technology is not a modern pursuit. The exact beginnings of the use of wind power are not agreed upon but some experts say that it was utilized as far back as the first century B.C. Historians concur that windmills were in existence as early as the seventh century. Up until the 18th century, windmills remained in use in various parts of the world, primarily for milling grain and pumping water.

Many new developments occurred in the 19th century. Wind power specialists in the United States and Denmark made great strides when they discovered how to harness wind power to generate electricity. Further advancements in this area occurred throughout the 20th century, with Sweden and Germany also making significant contributions. The future of wind technology is certain and trends indicate that offshore wind farms will become increasingly important in the 21st century.

KEY CONCEPTS AND TERMS

Dutch windmill
Hollow-post mill
Horizontal axis
Post windmill
Renewable energy source
Spring sail
Tower windmill
Vertical axis
Yawing

CHAPTER ASSESSMENT: HISTORY

1. The primary use of wind power technology in the 20th century was to mill grain.
   - ❑ **A.** True
   - ❑ **B.** False
2. Which innovator was responsible for creating wind turbines to generate electricity in rural Denmark in the early 20th century?
   - ❑ **A.** Palmer Cosslett Putnam
   - ❑ **B.** Poul La Cour
   - ❑ **C.** Ulrich Hütter
   - ❑ **D.** Daniel Halladay
3. What were the main uses of wind power prior to the 20th century? (Select three.)
   - ❑ **A.** Generating electricity
   - ❑ **B.** Pumping water
   - ❑ **C.** Herding cattle
   - ❑ **D.** Milling grain
   - ❑ **E.** Sailing ships

4. The Smith-Putnam wind turbine was constructed in:
   - ❑ A. the United States.
   - ❑ B. Germany.
   - ❑ C. Canada.
   - ❑ D. Denmark.
5. Which type of windmill incorporated a rotating tower cap that turned the wind wheel?
   - ❑ A. Post windmill
   - ❑ B. Dutch windmill
   - ❑ C. Tower windmill
   - ❑ D. Hollow-post windmill
   - ❑ E. Both B and D
6. Charles Brush of the United States was the first person to use wind power to generate electricity.
   - ❑ A. True
   - ❑ B. False
7. What is Ulrich Hütter of Germany well known for creating?
   - ❑ A. Spring sails
   - ❑ B. Europe's first offshore wind farm
   - ❑ C. Wind tunnel
   - ❑ D. W-34 wind turbine
   - ❑ E. 12 kW DC windmill generator
8. In 1984–1985 in California, the Pacific Gas and Electric Company (PG & E) and the Southern California Edison Company (SCE) covered costs up to ________ of the power channeled into their grids.
   - ❑ A. 5 percent
   - ❑ B. 10 percent
   - ❑ C. 15 percent
   - ❑ D. 25 percent
   - ❑ E. 0 percent
9. The California wind rush of the 1970s was spurred on by federal tax credits.
   - ❑ A. True
   - ❑ B. False
10. The energy crisis of the 1970s was caused by which of the following?
    - ❑ A. Lack of natural resources
    - ❑ B. Scarcity of crude oil
    - ❑ C. Increased costs of crude oil
    - ❑ D. All of the above

11. Apart from the United States, which countries led the way in using wind power technology in the 1970s? (Select three.)

- ❑ **A.** China
- ❑ **B.** Sweden
- ❑ **C.** Russia
- ❑ **D.** Denmark
- ❑ **E.** Germany

12. In the 1940s, Putnam outlined the technical data for the ideal wind turbine, which still applies today.

- ❑ **A.** True
- ❑ **B.** False

13. Which adjustments did Halladay make to windmills in the 19th century in the United States so that they would be self-regulating? (Select two.)

- ❑ **A.** He employed a rotating vertical axis.
- ❑ **B.** He added flyweights.
- ❑ **C.** He doubled the size of the windmill's foundation.
- ❑ **D.** He replaced sailcloth with a wind wheel made of wooden blades.

14. The ________ windmill typically had triangular shaped sails.

- ❑ **A.** post
- ❑ **B.** hollow-post
- ❑ **C.** tower
- ❑ **D.** Dutch

15. The popularity of wind turbines increased dramatically with the introduction of the US Rural Electrification Program in the 1930s.

- ❑ **A.** True
- ❑ **B.** False

chapter 2

Basic Concepts

AS THIS CHAPTER IS devoted to basic concepts, you are presented with the most elementary information, beginning with the physical parts that make up a wind tower. This introduction to a wind tower's components includes a look at the wind tower itself: blades, generators, and bearings. Additionally, an explanation of the different wind turbine size classes is included. As you review the information, do your best to visualize each component and how it fits into the wind turbine as a whole. Make use of the diagrams and photos to assist you in this process. Next, you are presented with the fundamentals of wind energy. You will see a few equations that require basic math skills. Take your time and get to know the new symbols and how they interact. Be sure you understand a concept before moving on to the next section. To finish the chapter, you will read about the significance of tower height.

Chapter Topics

This chapter covers the following topics and concepts:

- What components constitute a wind tower
- The basics of wind energy
- The importance of tower height

Chapter Goals

When you complete this chapter, you will be able to:

- Describe the different wind turbine size classes and the types of wind turbines found in each class
- Describe the various components that make up a wind tower
- Calculate power in the wind and explain the relevance of each variable in the equation
- Outline the importance of tower height

Wind Tower Components

Before looking at the basics of wind energy, it is important that you understand the various parts that make up a wind tower. In addition to a review of a wind tower's components, an outline of the different wind turbine size classes is necessary.

Wind Turbine Size Classes

When people think about the size of wind turbines, towering windmills usually come to mind. Contrary to popular belief, there are wind turbines small enough for one person to carry. Wind turbines are divided into two general categories: small and large. Within each of these groupings, there are three different sizes.

When describing the size of the wind turbine, what do you think is the most important factor? Some assume height is the determining factor. In fact, three factors are used to assess a wind turbine's size classification FIGURE 2-1. The first is **rotor diameter**. Note that the **rotor** includes both the blades and the hub. It is helpful to draw an imaginary circle around the blades, as is illustrated in the diagram below. Now imagine a line that cuts right through the middle of that circle one side to the other. The length of that line is called the rotor diameter. The second factor in assessing a turbine's size is **swept area**. This describes the amount of area that the blades cover or sweep. Swept area is always described in square feet (ft^2) or square meters (m^2). The third factor is **standard power rating**, which describes the amount of energy produced by the turbine. The primary factor that determines the size class of a wind turbine is the area swept by the rotor.

The use of small wind turbines in noncommercial environments is popular worldwide. **Distributed generation**, or incorporation of the use of wind power in populated zones, is gaining in popularity in North America. Europe has been employing small wind turbines in and around villages and cities with success.

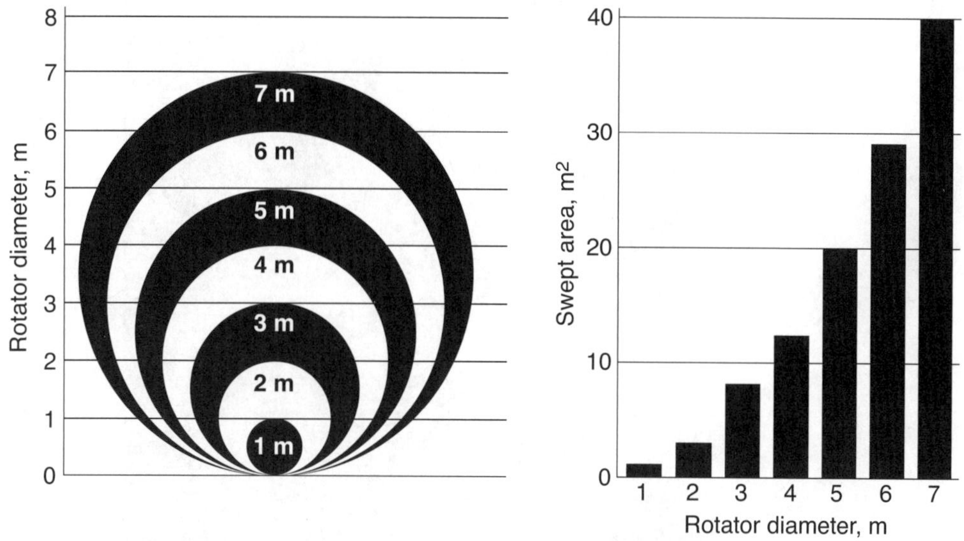

FIGURE 2-1 The basics of a wind turbine: The rotor diameter and the swept area are two of the three factors used to assess the size of a wind turbine.

Increasingly, small wind turbines are also providing energy on sailboats, at secluded residences, and in dwellings off the grid.

The following outlines the three types of small wind turbines:

- **Micro**—These are the smallest types of wind turbines; they have a rotor diameter of 2 to 4 ft (0.2 to 1.2 m) and a swept area of 2 to 13 ft^2. Their standard power rating ranges from 0.04 kW up to 0.25 kW. Two examples of common micro wind turbines are the 200-watt turbine called Air Breeze made by Southwest Windpower and Model 300 (shown in **FIGURE 2-2**) made by Ampair. Both have a swept area or intercept of approximately 1 square meter. These turbines are typically used for recreational purposes, such as on sailboats and for RVs (recreational vehicles).
- **Mini**—These wind turbines are a bit larger than micro wind turbines. The average rotor diameter runs between 4 to 10 feet (1.25 to 3 m) with a swept area of 13 to 76 ft^2 (1.2 to 7.1 m^2). Common mini wind turbines include the Whisper 100 (shown in **FIGURE 2-3**) by Southwest Windpower, with a swept area of 3.6 m^2, and the Bergey Windpower's XL.1, which has a swept area of 4.9 m^2. In Britain, you can see mini wind turbines set up on the rooftops of buildings in the cities.
- **Household**—These wind turbines are classified as the biggest type of small wind turbine. The range within this category is wide. The standard power

FIGURE 2-2 The Ampair 300 is an example of a micro wind turbine.

© Realimage/Alamy Images

FIGURE 2-3 The Whisper H40, a mini turbine now called the Whisper 100, is more than four times the size of a micro wind turbine.

rating for this type of turbine is between 1.4 and 16 kW. The rotor diameter averages 10 to 33 feet and the swept area ranges from 76 to 845 ft^2 (7–79 m^2). Examples include Southwest Windpower's Skystream, which has a swept area of 10 m^2, and the Bergey Excel with a swept area of almost 40 m^2.

Historically, large wind turbines have predominated. Today, their uses range from farms to power plants. There are three categories of large wind turbines. They are:

- **Small commercial**—These wind turbines are the smallest in this category. The rotor diameter can be as little as 33 feet (10 meters) and as much as 66 feet (20 meters). The swept area for this type of large wind turbine ranges between 840 and 3,400 ft^2 (79–314 m^2). The power produced runs between 25 and 100 kW. These types of turbines are typically used for farms, schools, and small businesses. One example is Entegrity's EW50, which has a swept area of 175 m^2.
- **Medium commercial**—The rotor diameter of medium-sized turbines range between 66 and 164 ft (20–50 m) and the swept area is 3,400 to 21,100 ft^2 (314 to 1,963 m^2). The typical power rating for these large wind turbines falls between 100 and 1,000 kW. Most commonly, these turbines are used for farms, factories, businesses, and small wind farms. In the capital of New Zealand, a medium-sized commercial turbine, the Vestas V27 (shown in FIGURE 2-4), is one of the world's most efficient turbines.
- **Large commercial**—The largest of all the wind turbines have a rotor diameter that ranges in size from 164 to 328 ft (50 to 100m) and a swept area of 21,100 to 84,500 ft^2 (1,963 to 7,854 m^2). Their power ratings fall between 1,000 and 3,000 kW. These wind turbines are most commonly used in modern wind power plants. However, these turbines can also be used on their own. An example of this type of turbine is Sky Generation's Vestas V80 (shown in FIGURE 2-5), which has a swept area of more than 5,000 m^2.

Towers

All components of a wind turbine are important but obviously the tower is essential. It is the mechanism that raises the turbine so that it can access the wind. Starting at the bottom of every wind tower is the tower **base**, usually made of concrete. The base sits in a hole and this is where the tower tube or legs of the tower rest. For maximum stability, every hole must be dug below the **frost line**. If a tower is built above the frost line, frost can push the soil and the tower out of the ground. Within the concrete base you also find metal rods that provide additional strength to the base, known as **rebar**.

In addition to the base, some wind towers also have what is called an **anchor**. An anchor, like the base, is also made of concrete, sits in a hole, and contains

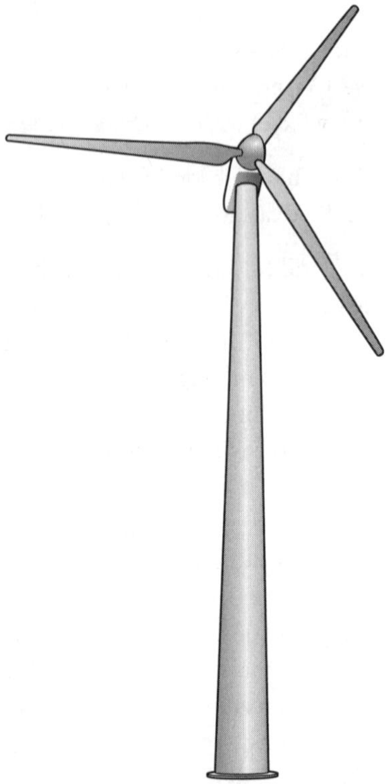

FIGURE 2-4 The Vestas V27 is a medium-sized commercial turbine currently in use in Wellington, New Zealand.

rebar. Additionally, steel is embedded in the concrete and extends out to connect with metal cables known as **guy wires**. Some towers have three anchors and others have four. The anchor, just like the anchor of a ship, prevents the tower from moving. But anchors are necessary only when the tower relies on guy wires. For some wind towers, such as a fixed guyed or a tilt-up tower, guy wires are essential to stabilize the tower. Freestanding towers do not require guy wires. Guy wires come in two different types. One is a cable made of a flexible material and the other is called a **guy strand**, which is not flexible and quite durable. Towers almost always use steel as their primary material.

One last component of wind towers is the **transmission wires**. Without the wires, the electricity generated by the wind would not reach the ground. They are found strapped along the tower starting from the top of the tower where the wind generator is located. Sometimes they run inside the tower tube. They continue down through the base and underground until they reach the electronics room.

FIGURE 2-5 The Vestas V80, located in the province of Ontario in Canada, is an example of a large commercial turbine.

A special casing made of plastic or metal called **conduit** protects the transmission wires.

Types of Propellers

Most people assume that wind turbines always have three or four propellers or blades. In fact, the number of blades varies from turbine to turbine. What many people don't realize is that only one blade is needed to transform energy from the wind. One-blade turbines were used in the past but they are no longer sold today.

Wind turbines can have anywhere from two to five blades, but three-blade turbines are the most common type. Generally, three-blade turbines function more smoothly and last longer than two-blade turbines. This is reflected in the turbine market in that only a handful of manufacturers of large wind turbines still produce two-blade turbines. Four-blade turbines have existed, but like one-blade turbines, they are no longer on the market for large commercial turbines. Turbines with more than four blades also exist.

Almost all blades today are composed of a mix of fiberglass, carbon fiber, or wood, which are the most lightweight and durable materials. In the past, blades have been made out of heavier and less durable materials, such as cloth, steel, and aluminum.

Electrical Generators

You can think of the generator as the heart of the turbine. The generator is found at the core of a larger electrical system. It is the generator's job to convert the wind or mechanical energy into electricity. Generators can produce two types of electrical current:

- **Direct current (DC)**—A flow of electricity that moves in one direction and remains consistent. Generators that produce a direct current are costly and require much maintenance, so they are not commonly used. The exception is very small wind turbines that recharge batteries.
- **Alternating current (AC)**—A flow of electricity that moves back and forth in two opposing directions. Most turbines produce an alternating current.

Electrical generators use magnetic fields to transform rotational energy into electrical energy. There are two basic types of electrical generators used for wind turbines, which are:

- **Synchronous generator**—Current is generated by the rotating movement of the rotor while coils move magnets at the same speed as the alternating current. Synchronous refers to the identical speed of the rotor and the magnets.
- **Asynchronous or induction generator**—Current is also generated by a magnetic field. This field is created by the rotating movement of the rotor and a rotating magnetic field induced in the **stator** or the stationary part of the rotor. It is called asynchronous because the speed of the rotor and the speed of the magnetic field created by the stator are not the same.

A few designers prefer the use of wound-field generators, which strengthen magnetic fields through rotation of the rotor. Wound-field generators can operate at variable speeds as well, which helps ensure long life for the turbine.

If a generator relies on a battery, the following two mechanisms should also be present:

- **Charge controller**—Acts to shield the batteries from working too hard and thus losing their charge
- **Inverter**—Transforms a direct current (DC) into an alternating current (AC)

Bearings

Another basic component of wind turbines is bearings. Bearings are the areas that support the friction caused by a rotating object. In wind turbines, **yaw bearings**,

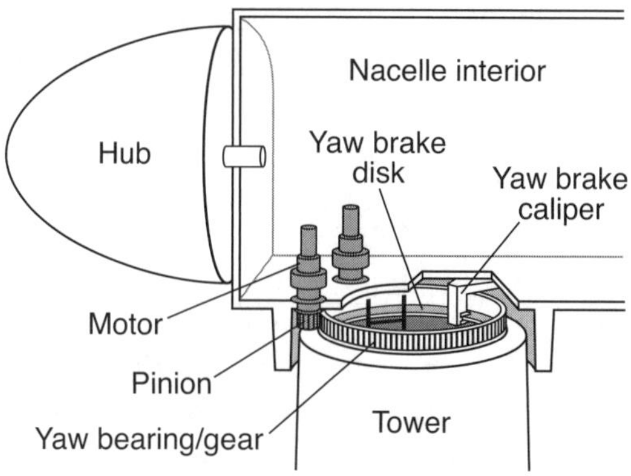

FIGURE 2-6 The elementary components of a wind turbine.

shown in **FIGURE 2-6**, play an important role. If you watch a wind turbine in action, notice that the head of the turbine follows the wind. In fact, it can turn 360 degrees, thanks to the yaw bearings.

Wind Energy Basics

Getting down to the basics of wind energy means knowing how to calculate the power in the wind. But before you do any calculations, you need an explanation of the variables involved. This section ends with a review of two elementary but essential equations.

Power

Would you describe power (P) as being the same as energy? From a scientific perspective, they are two different things. When power is used in a technical sense it refers to **watts** or wattage, which measures the rate of energy generation, movement, or use. In contrast, energy is referred to as **watt-hours**, which measure electrical units. Also, be aware that power is a measure of energy in the moment, not the total of energy gathered over a specific time.

Air Density

An essential fact related to **air density (ρ)** is that more power is available in dense air than thin air. But air density is a variable that does not experience great changes. In fact, changes occur only depending on the environment. Two environmental factors have the power to affect air density: temperature and elevation.

Temperature and air density have an inverse, or opposing, relationship. This means that as temperature drops, air density rises. In order to access more power, setting up a wind turbine in a cooler climate is recommended. Elevation and air

> **NOTE**
>
> For every 1,000 feet above sea level, a turbine's approximate performance drops 3 percent.

density also have an inverse relationship in that as elevation increases, air density drops. Therefore, sea level has the densest air when compared to air at other elevations. Both factors have to be considered when assessing the potential performance of a wind turbine.

Swept Area

Recall that swept area (A) describes the circular area that the rotor blades of a turbine cover. Think of this as the area that captures or collects the wind. The blades contain the wind energy, which is then channeled behind the rotor into the turbine's generator. That is where electricity is produced. Swept area (with the rotor diameter) is considered the most important predictor of a wind turbine's potential. Imagine two turbines side by side. The only difference is that turbine A has a rotor that is two times the size of turbine B. The other variables of air density and wind speed are the same. Ultimately, turbine A is predicted to have two times as much power potential.

To determine swept area, use the following equation:

$$A = \pi r^2$$

$$A = 3.14 \text{ (pi)} \times \text{circle's radius (half the circle's diameter)} \times \text{circle's radius}$$

Assume that a turbine has a 4-foot diameter. Half of the diameter, 2 feet, determines the radius. Look at the equation:

$$A = 3.14 \times 2 \times 2$$

$$A = 12.5 \text{ square feet (ft}^2\text{)}$$

Examine the difference in swept area if the diameter was doubled to 8 feet:

$$A = 3.14 \times 4 \times 4$$

$$A = 50 \text{ ft}^2$$

> **NOTE**
>
> You must know the diameter of the area that the rotor occupies to calculate swept area.

By doubling the rotor diameter, the swept area is approximately four times as large, which greatly increases its power potential. This explains why many wind technology specialists measure and rank turbines according to their swept area.

Cube of Wind Speed

Remember that wind speed (V) does not provide an accurate reflection of the power in the wind. Rather, it is when wind speed is cubed (V^3) or multiplied three times ($V \times V \times V$) that we see its true potential.

> **NOTE**
>
> Wind speed is represented by the capital letter *V*. *V* stands for velocity, a synonym of speed.

Assume the speed is 5 miles per hour (mph); $5 \times 5 \times 5 = 125$ mph. Double that to 10 mph and the wind speed cubed would be $10 \times 10 \times 10 = 1{,}000$. At 10 mph, wind speed cubed is much more than the double of 5 mph wind speed cubed.

Height and Wind Speed

Although you can't change wind speed, you can access areas that have increased wind speeds. One of the best ways to capture stronger winds is by increasing the height of the tower. To determine the exact increase in wind speed with height, you turn to the power law method. It looks like this:

$$V = (H/H_0)^{\alpha} V_0$$

$$\text{Wind speed at new height} = (\text{new height/original height}_0)^{\text{surface roughness}} \times \text{wind speed at original height}_0$$

The surface roughness variable or α relates to the impact plant life and landscape have on wind speed. A rough terrain can cause increases in wind speed, sometimes as high as 0.4. By comparison, a flat landscape would have a surface roughness variable no higher than 0.1. Combining this exponential variable with tower height can substantially increase power in the wind.

Researchers believe that wind speed power is maximized when the distance between the top of obstructions, such as treetops, and the tip of a turbine increases. A minimum of 10 meters above an obstruction is necessary to yield effects.

Power in the Wind Equation

To calculate the amount of power in the wind, you first need to understand what each symbol stands for:

- *P*—power in the wind
- *D* or ρ—air density
- *A*—area intercepting the wind or swept area
- *V*—speed

The calculate the power in the wind, use the following equation:

$$P = \tfrac{1}{2}\,\rho A V^3$$

$$\text{Power in the wind} = \tfrac{1}{2} \times \text{air density} \times \text{intercept area/swept area} \times \text{speed cubed}$$

As you can see from the equation, if there is an increase in air density, the intercept area, or the speed of the wind, more power from the wind can be accessed. Wind speed is especially important. However, because wind speed is cubed, even the smallest increase in wind speed can have grand effects on the potential power. For instance, look at the math for wind speed that has tripled:

$$(3^3) = 3 \times 3 \times 3 = 27$$

That means there is 27 times as much power available, not just 3 times as much.

> **NOTE**
>
> Notice that air density can be represented by *D* or ρ, but most commonly you will see it as ρ.

Air Density Equation

Take note that when air density in kg/m^3 at sea level replaces ρ, power in watts is determined by the following equation:

$$P = \frac{1}{2}(1.225)AV^3$$

Multiplying 1.225 and ½ gives 0.6125. Therefore, the final equation is:

$$P = 0.6125\,AV^3$$

Power in the wind = 0.6125 × area in square meters (m^2) × speed in meters per second (m/s)

TOWER HEIGHT

Imagine your very own wind turbine. If you had the choice of obtaining a larger generator or a taller tower, which would you choose to maximize the efficiency of your wind turbine? Experts agree that you should opt for a taller tower. Consider that by doubling the height of a tower you can access up to 25 percent more power. Make a tower five times its height and you double the available power. Lack of elevation is one of the primary reasons wind turbines fail to operate. Wind specialists strongly advise against installing a turbine on the roof of your home. Chances are that trees and other homes would be nearby. These act as obstacles to the wind and create turbulence, which reduces wind speed. It would not collect significant power from the wind simply because it was not tall enough.

The trend for taller and taller towers began recently. Through the 1960s up until the 1990s wind towers usually reached a cap of 50 meters. By the late 1990s, tower heights reached as high as 70 meters. The first decade of the 21st century witnessed much growth in technology and design. With rotor diameters steadily increasing in size, tower heights had to go up as well. After the year 2000, an average hub height was 100 meters. Then in 2009, one of the highest towers in the world was constructed in Germany, shown in **FIGURE 2-7**. It sat at an incredible 133 meters. In comparison to a 100-meter tower, the 133-meter tower could produce up to 20 percent more energy.

More people are realizing the greater efficiency of tall towers. In Cornwall, England, the Delabole wind farm has existed since 1991, originally under private ownership. In 2002, the company Good Energy took over. Using 10 turbines with a hub height of 50 meters, the company generated power for approximately 2,700 families. More recently, however, those 10 turbines were replaced by four. The four new turbines reach a height of 99 meters, nearly double their original height. The company says that the total combined capacity is now 9.2 MW, enough to supply 7,000 homes.

Despite the advantages of taller towers, wind turbine manufacturers have discovered that increased height brings its own set of problems. One component that can be affected is the foundation.

TOWER HEIGHT (Continued)

FIGURE 2-7 ATS Projekt Grevenbroich GmbH built and ran the 133-meter tower in Grevenbroich, Germany.

The added height can cause cracks in the foundation. Additionally, the turbine can also be affected. With increased wind speed at higher heights, the strain on the turbine is intense. Wind speeds over 50 mph can sometimes cause the turbine to go into overdrive. This can lead to a complete shutdown. Blades also receive more wear and tear at increased heights, which can lead to ineffective performance. Although tall towers have their fair share of challenges, engineers are addressing these issues so that the benefits of tall towers will outweigh the costs.

CHAPTER SUMMARY

You should now be clear on the most important components of a wind tower. Beyond examining wind tower parts and wind turbine size classes, you were introduced to the basics of wind energy. At this point, be sure you grasp the concepts of power, air density, wind speed cubed, swept area, and the effect of height on wind speed. Anyone can plug numbers into an equation. What counts is that you comprehend how each variable in the equation interacts with the others.

KEY CONCEPTS AND TERMS

Air density
Anchor
Base
Charge controller
Conduit
Distributed generation
Frost line
Guy strand
Guy wires
Inverter
Rebar
Rotor
Rotor diameter
Standard power rating
Stator
Swept area
Transmission wires
Watt (W)
Watt-hours (Wh)
Yaw bearings

CHAPTER ASSESSMENT: BASIC CONCEPTS

1. Three-blade turbines are the most popular type of turbine today.
- ❑ **A.** True
- ❑ **B.** False

2. Which is the equation to calculate swept area?
- ❑ **A.** $A = \pi r^3$
- ❑ **B.** $A = \pi^2 r$
- ❑ **C.** $A = \pi^3 r$
- ❑ **D.** $A = \pi r^2$

3. Swept area is directly related to:
- ❑ **A.** wind speed.
- ❑ **B.** rotor diameter.
- ❑ **C.** temperature.
- ❑ **D.** Both B and C

4. Which type of wind turbine is most commonly used on farms? (Select two.)
- ❑ A. Micro
- ❑ B. Mini
- ❑ C. Small commercial
- ❑ D. Medium commercial
- ❑ E. Large commercial

5. Which is the correct equation to calculate the power in the wind?
- ❑ A. $P = ½\ \rho DV^3$
- ❑ B. $P = ½\ \pi AV^3$
- ❑ C. $P = ½\ \alpha AV^3$
- ❑ D. $P = ½\ \alpha DV^3$
- ❑ E. None of the above

6. Which of the following is the most important factor in determining the classification of a wind turbine?
- ❑ A. Rotor diameter
- ❑ B. Tower height
- ❑ C. Generator size
- ❑ D. Access to high winds

7. The value that represents wind speed is multiplied three times in the power equation.
- ❑ A. True
- ❑ B. False

8. Which two factors affect air density? (Select two.)
- ❑ A. Elevation
- ❑ B. Obstructions
- ❑ C. Surface roughness
- ❑ D. Temperature

9. Power measures the accumulation of energy over time.
- ❑ A. True
- ❑ B. False

10. Which type of wind turbine is typically used on sailboats or recreational vehicles?
- ❑ A. Micro
- ❑ B. Mini
- ❑ C. Household
- ❑ D. None of the above

11. The ________ allow the head of the turbine to turn 360 degrees.
- ❑ A. yaw bearings
- ❑ B. guy wires
- ❑ C. transmission wires
- ❑ D. guy strands

12. Where is the rebar located on a wind tower? (Select two.)
- ❑ **A.** Anchor
- ❑ **B.** Generator
- ❑ **C.** Base
- ❑ **D.** Rotor
- ❑ **E.** Tower

13. What does the ρ symbol stand for?
- ❑ **A.** Surface roughness
- ❑ **B.** Velocity
- ❑ **C.** Power
- ❑ **D.** Air density
- ❑ **E.** None of the above

14. Technically speaking, ________ refers to watts or wattage.
- ❑ **A.** Velocity
- ❑ **B.** Electrical units
- ❑ **C.** Power
- ❑ **D.** Energy

15. Rotor blades are most commonly composed of fiberglass and aluminum.
- ❑ **A.** True
- ❑ **B.** False

Wind Energy Converters

chapter

3

THIS CHAPTER IS DEVOTED to explaining the basics of wind energy converters, concentrating on the role of the rotor. Take note that the rotor is not solely responsible for the conversion of wind to power but it plays the largest role in wind energy conversion. The rotor includes both the turbine's hub and its blades. A review of the two basic types of wind energy converters is explored as well as an explanation of how a wind energy converter's power yield can be augmented. The chapter concludes with information on the growing trend of community wind farms.

Chapter Topics

This chapter covers the following topics and concepts:

- Types of wind energy converters
- How a wind energy converter's power yield can be increased
- Community wind and its benefits

Chapter Goals

When you complete this chapter, you will be able to:

- Describe and differentiate between the two major types of wind energy converters
- Outline the various ways that the power yield of a wind energy converter can be augmented
- Explain the concept of community wind and its benefits

Types of Wind Energy Converters

In the field of wind technology, experts classify wind energy converters by looking at the rotor and how it is positioned. More specifically, they use the axis of rotation or the angle of the rotor to differentiate wind energy converters. Two different angles are used for wind turbines. One uses a vertical axis that rotates at a right angle to the horizon. The other uses a horizontal axis that rotates parallel to the ground. Understand that although wind energy converters are categorized according to their physical design, other parts of a wind turbine, such as the **generator** and the **gearbox**, also play a role in the process of converting mechanical energy to electrical energy.

Vertical Axis Rotors

The vertical axis design of rotors is the oldest but is currently less common. You may also see this type referred to as vertical axis wind turbines (VAWTs). One advantage of rotors with a vertical axis of rotation is that its design is not complex. Another plus of this design is that it allows for the option of easily containing the gearbox and the generator on the ground, instead of behind the rotor. Furthermore, these wind energy converters do not require a **yaw system**, which controls the rotor so that it faces the wind. On the other hand, rotors with a vertical axis of rotation have their disadvantages. First, they do not contain a self-starting function. Second, most do not possess blade pitch control, which can control power output and speed; but newer designs do include more sophisticated blade shape and pitch control. Lastly, they have a low **tip-speed ratio**, which is defined as the ratio between the rotational speed of the tip of the blade and wind speed. A high tip-speed ratio indicates increased efficiency.

Among this type, a few designs have been used. An earlier design from 1925 (see FIGURE 3-1), called the Darrieus rotor, named after its creator, shows potential today because it can generate what is called aerodynamic lift, which is a mechanical force caused by the motion of the rotor. It usually is built with two or three thin curved blades. The rotating blade is best described by its Greek name of *troposkien*, which means "turning rope." A downside to this type is its complex blade design, which makes producing it a difficult task.

Another type of rotor with a vertical axis of rotation is the H-rotor, which is an adaptation of the Darrrieus rotor. The primary difference between the two models is the shape of the blades. The H-rotor used straight blades as opposed to curved ones. Although a gearbox is not needed with this design, it still carries a high cost to manufacture. A different VAWT design is called the Savonius. The blades are in the form of scoops and the rotor, when looked at from an aerial view, is shaped like the letter 's.' Because of its low tip-speed ratio this design isn't used to generate electricity. Most commonly it is used in small wind rotors that drive small water pumps. FIGURE 3-2 shows you the primary types of rotors that operate on a vertical axis of rotation.

FIGURE 3-1 An example of a wind turbine using the Darrieus rotor.
Courtesy of aarchiba

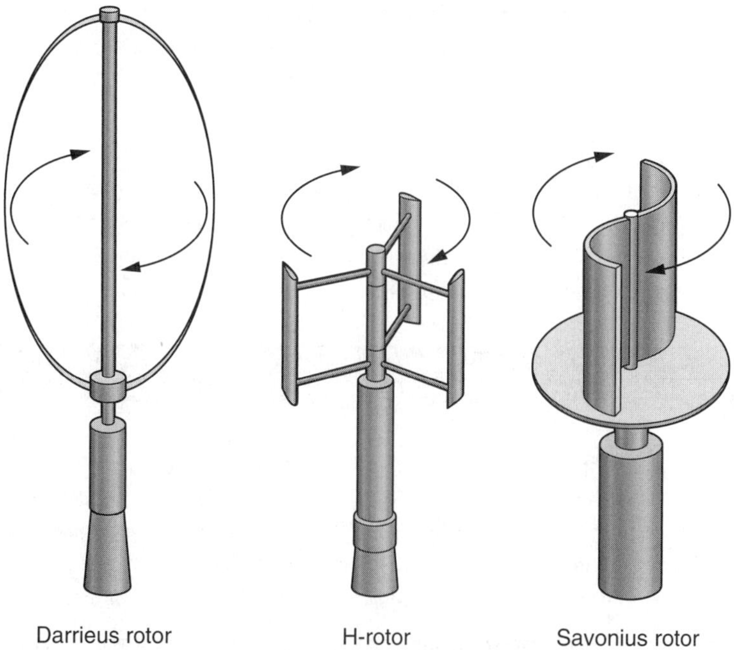

FIGURE 3-2 The three primary designs of rotors that operate on a vertical axis.
Adapted from Hau, E. (2006). *Wind Turbines.* Berlin and Heidelberg: Springer-Verlag.

Horizontal Axis Rotors

You are probably most familiar with wind energy converters that use a horizontal rotation. At present, horizontal axis rotors with blades that resemble propellers are the most common type of wind energy converters. They can also be referred to as horizontal axis wind turbines (HAWTs), shown in **FIGURE 3-3**. The advantages of using this design are threefold. First, unlike converters that employ a vertical rotation, rotors on a horizontal axis are equipped with blade pitch control. Both the speed of the rotor and the power output can be managed with this capacity. Next, the shape of the rotor blades can increase the power output. Lastly, technology is continually advancing in the realm of propeller design and new and improved designs are forthcoming. Take note that in contrast to VAWTs, the generator is found behind the rotor in HAWTs.

Increasing Power Yield

Concentrators play a key role in increasing the efficient functioning of the rotor. A **concentrator** acts to enhance wind energy by concentrating it or condensing it, which ultimately increases rotor efficiency. If you increase rotor efficiency, you also increase power yield. Wind technology is constantly evolving and new concepts are continually being tested. A review of six different concepts will reveal the variety and the ingenuity of concentrators. Before this review however, a brief discussion of Albert Betz is necessary.

Albert Betz is renowned in the field of wind technology for discovering the **power coefficient**, also written as C_p. This term tells you how well a wind turbine converts wind to electricity. More specifically, in the 1920s Betz identified the maximum fraction of power than can be garnered from the wind, which is written as $C_p = 0.593$. In other words, a maximum 59.3 percent of wind energy can be converted into power. The C_p can also be referred to as the Betz value or the Betz limit. An explanation of the equations behind Betz's power coefficient is not necessary at this time. However, know that when discussing concentrators and ways to increase power yield, the Betz value is a used as a measuring stick for assessing rotor efficiency.

BUYER BEWARE AND BETZ LAW

If you are shopping for a wind turbine and a manufacturer promises a turbine that has a power coefficient greater than the Betz limit of 0.593, don't be fooled. Often, more is better for the consumer but not in the case of wind energy. The Betz limit, also called the Betz Law, named after the German physicist who discovered the value almost a century ago, shows the maximum amount of kinetic energy that can be derived from the wind. The maximum—about

(Continues)

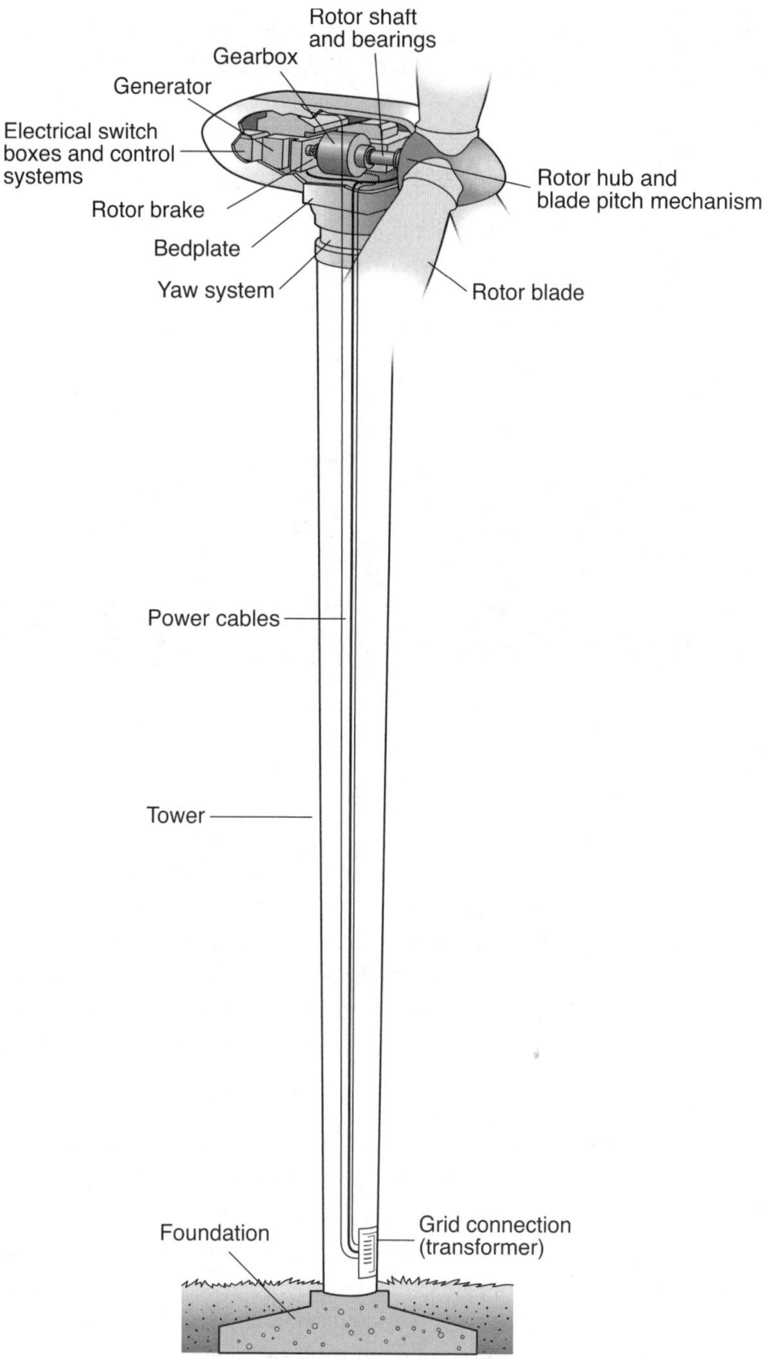

FIGURE 3-3 The parts that make up a HAWT.

Adapted from Hau, E. (2006). *Wind Turbines*. Berlin and Heidelberg: Springer-Verlag.

BUYER BEWARE AND BETZ LAW (Continued)

59 percent—occurs under ideal conditions. Therefore, if wind speeds are low or the turbine is not functioning at full efficiency, the amount of energy accessed by the wind will be less than 59 percent.

But this isn't necessarily a bad thing. If 100 percent of wind energy is accessed and converted into kinetic energy, what will propel the turbine's blades? Utilizing all the wind energy would create stillness, an obvious obstacle to running a turbine. In other words, for maximum performance, a turbine should not utilize all the available wind.

A few manufacturers can now claim that wind power extraction can go beyond the Betz limit without contradicting the Betz Law. This is achieved by a device that captures heat from passing wind. (See below in "Rotors Using Ducts.") This theory was first presented by Nicola Tesla, a 20th-century mechanical engineer.

The bottom line? Buyer beware. If you come across "state of the art" turbines that advertise a power coefficient greater than the Betz value, you can bet that behind those claims are problematic calculations and testing conditions. In the world of wind turbines, less is more.

Rotors Using Ducts

One strategy for increasing rotor efficiency is to house the rotor in a duct, a hollow tube that acts as a channel. Without a duct, power is lost in the free air stream. A duct prevents some loss, which means more power. This type of duct has a power coefficient greater than the Betz value: $C_p = 0.66$. Another type of concentrator is a diffuser duct (see **FIGURE 3-4**), which is essentially shaped like a reverse funnel. The diffuser acts to increase the flow of the wind stream. This acts to moderately increase the power yield; it has a power coefficient of about 0.75.

Vortex Tower

It should be noted that this type of concentrator has not yet been applied to an actual wind turbine. To date it has only been tested in a wind tunnel. As shown in **FIGURE 3-5**, the concept focuses on creating a vortex or a whirlpool of air, like a tornado, on top of the wind flow. The vortex or tornado tower is built with shutters on the exterior of a cylindrical container. The wind that flows through it creates a vortex, thus feeding the rotor with additional force. It doesn't appear likely that this concept will be readily applied to wind turbines. In addition to predictions that it would create unacceptable noise levels, its power coefficient sits at a low 0.1.

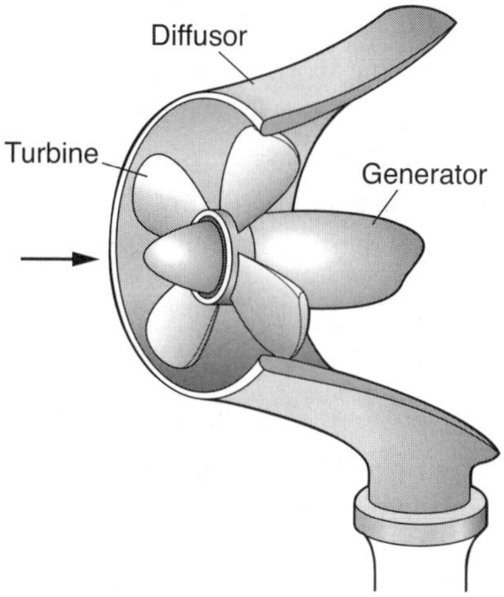

FIGURE 3-4 A wind turbine that incorporates a diffuser.
Adapted from Hau, E. (2006). *Wind Turbines*. Berlin and Heidelberg: Springer-Verlag.

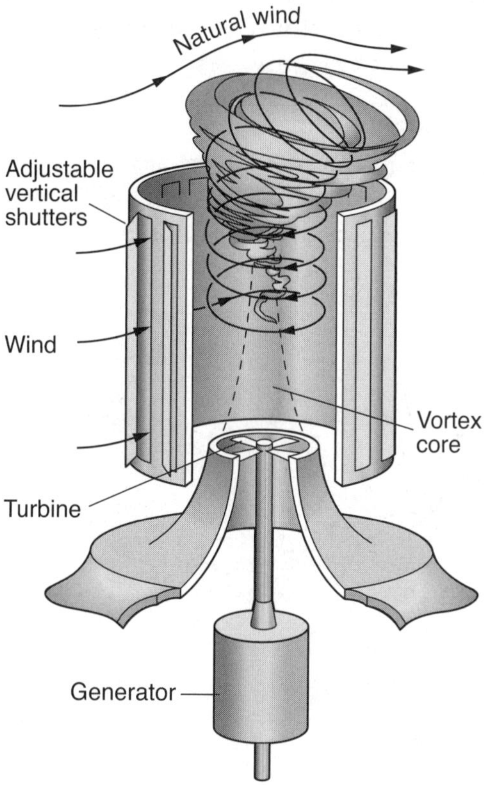

FIGURE 3-5 An example of a vortex tower.
Adapted from Hau, E. (2006). *Wind Turbines*. Berlin and Heidelberg: Springer-Verlag.

Vortex Concentration Using a Delta Wing

Scientists discovered that airplanes with large angled wings, also known as delta wings, create various air vortices in the flow along the boundary of the wing. In an attempt to create a more efficient rotor by using this natural occurrence, rotors were placed alongside each other on a structure that resembles the wing of aircraft **FIGURE 3-6**. Unfortunately, trial runs in a wind tunnel produced poor results.

Berwain Concentrator

The Berwain concentrator, shown in **FIGURE 3-7**, consists of a fixed structure in the shape of a wheel called a stator, which is equipped with various blades. A small wind rotor sits directly in the middle of the structure. The result is a powerful vortex that allows substantially more wind power to be accessed. This concentrator has been tested in controlled settings and in the field confirming an increase in accessible wind power. However, the fixed wheel stator needs to have movable blades in order to withstand severe winds. The engineering and costs involved are extensive. Newer variations on this design have been made at very low costs using existing water towers and other structures, often implemented in third world countries using scrap as source materials.

Thermal Upwind Model

The concept of thermal upwind relies on the fact that airflow is created with increases in temperature. As you can see in **FIGURE 3-8**, radiation from the sun, or solar radiation, is also incorporated into this concept. A tall tower is covered

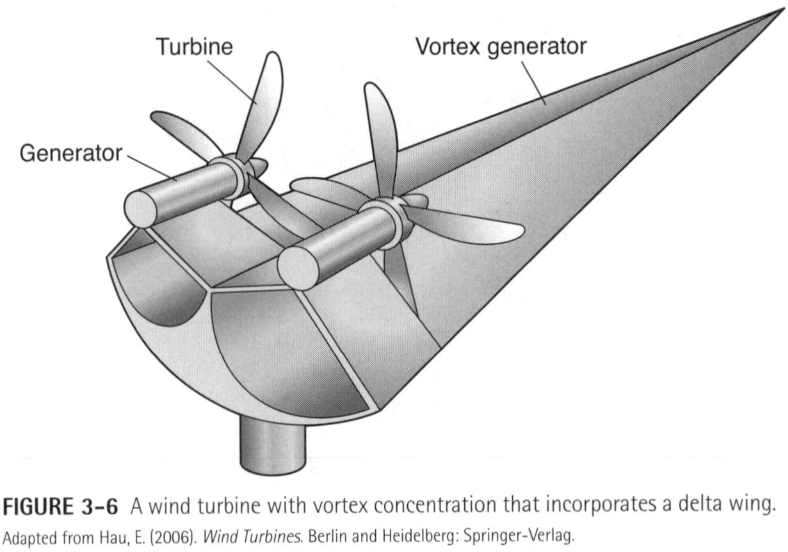

FIGURE 3-6 A wind turbine with vortex concentration that incorporates a delta wing.
Adapted from Hau, E. (2006). *Wind Turbines.* Berlin and Heidelberg: Springer-Verlag.

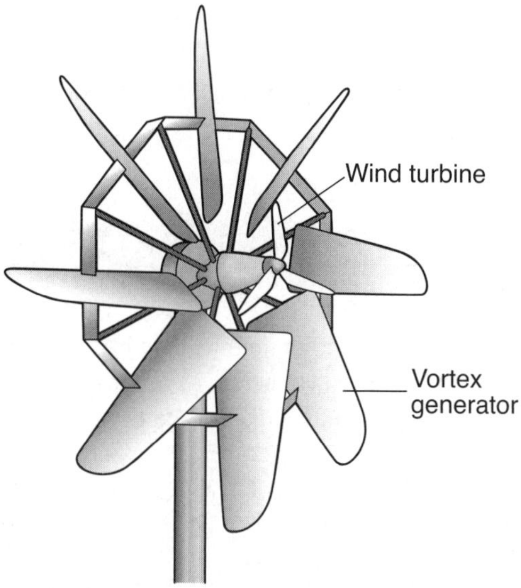

FIGURE 3-7 This drawing illustrates the concentrator or Berwain wind turbine.
Adapted from Hau, E. (2006). *Wind Turbines*. Berlin and Heidelberg: Springer-Verlag.

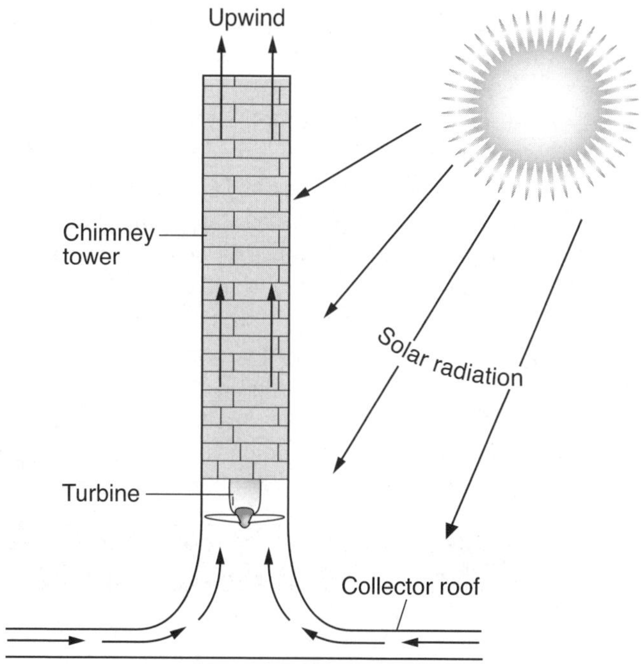

FIGURE 3-8 Illustration of the thermal upwind concept.
Adapted from Hau, E. (2006). *Wind Turbines*. Berlin and Heidelberg: Springer-Verlag.

with a collector roof designed to absorb the radiation from the sun. A wind that travels upward is created inside the tower, which provides energy to the turbine within the tower. Researchers have determined that this design functions best on a large scale.

WHAT IS COMMUNITY WIND?

When groups and individuals from a local community, such as businesses, schools, farmers, municipal utility companies, and cooperatives collectively invest in a wind energy project it is referred to as community wind. A wind energy project can consist of one wind turbine or a farm with various wind turbines. Usually, community wind turbines produce between 5 and 20 MW but lower or higher power yields are possible. In 2008 in the United States, community wind constituted approximately 4 percent of all of the functioning wind turbines across the country.

The trend is gaining in popularity for a variety of reasons. One of the most positive outcomes of community wind is the strengthening of the local economy. This can occur in various ways. First, the use of local labor and resources stimulates the community's economy by creating jobs. Next, profits from successful projects supply dividend income to the owners. As well, if local banks provide loans to the project the potential for positive financial impact exists. Furthermore, energy prices become stable because costs for transporting or mining energy are eliminated. In addition to the economic benefits, the community has much autonomy over how the project runs which contributes to a more unified collective.

One example of a successful project using this alternative form is found in northwest Ohio, where more than 1,800 families receive their electricity from community wind. In 2003, the City of Bowling Green, in addition to nine other local communities, became part owners of the Bowling Green Wind Farm, the first community wind project in the state. Many other states are following Ohio's example. For instance, the State of Washington is providing the Cascade Community Wind Company with $1 million to develop community wind. About one third of that funding comes in the form of a grant and over two thirds is given as a low-interest loan. Two community wind turbines are already up and running and several more are planned. Another recent example is a community wind project in Minnesota. A wind farm with a power yield of 5 MW will be developed with Juhl Wind and Gundersen Lutheran Hospital.

Developing community wind farms take funding, coordination, and commitment; but as many projects throughout the United States have realized, it is worth the effort and investment.

CHAPTER SUMMARY

You now know that a wind turbine's rotor plays a significant role in converting wind to power. The two basic types of wind converters are classified according to their design. Either they operate on a horizontal axis or a vertical axis. Remember that the most popular type today is the horizontal axis wind turbine or HAWT. An exploration of concentrators revealed the continual attempts to construct more efficient and cost-efficient turbines. Lastly, the advantages to using community wind are plenty and explain why community wind farms are becoming increasingly popular today.

KEY CONCEPTS AND TERMS

Concentrator
Gearbox
Generator
Power coefficient
Tip-speed ratio
Yaw system

CHAPTER ASSESSMENT: WIND ENERGY CONVERTERS

1. Vortex concentration (with or without a delta wing) is effective in increasing power yield.
   - ❑ A. True
   - ❑ B. False
2. Where is the generator located in a horizontal axis wind turbine?
   - ❑ A. Alongside the hub
   - ❑ B. Behind the rotor
   - ❑ C. On the ground
   - ❑ D. Inside the tower
3. What are the two main challenges of the Berwain concentrator? (Select two.)
   - ❑ A. Low power coefficient
   - ❑ B. Its ability to withstand extreme winds
   - ❑ C. The location of the generator
   - ❑ D. Extensive costs
4. Albert Betz determined that the ideal power coefficient is:
   - ❑ A. 0.603.
   - ❑ B. 0.593.
   - ❑ C. 0.583.
   - ❑ D. 0.589.

5. What are three benefits of community wind? (Select three.)
   - ❑ **A.** Guaranteed subsidies from local and federal governments
   - ❑ **B.** Generation of local jobs
   - ❑ **C.** Stability of energy prices
   - ❑ **D.** Elimination of transportation costs
   - ❑ **E.** Encouragement of international investments
6. The thermal upwind model incorporates solar radiation.
   - ❑ **A.** True
   - ❑ **B.** False
7. Which type of VAWT uses blades that form the shape of an 's'?
   - ❑ **A.** Darrieus H-rotor
   - ❑ **B.** H-rotor
   - ❑ **C.** Darrieus rotor
   - ❑ **D.** Savonius rotor
8. Vertical axis wind turbines do not require a:
   - ❑ **A.** yaw system.
   - ❑ **B.** generator.
   - ❑ **C.** gearbox.
   - ❑ **D.** stator.
   - ❑ **E.** hub.
9. VAWTs are the most common wind energy converter in use today.
   - ❑ **A.** True
   - ❑ **B.** False
10. Which of the following does a horizontal axis wind turbine have that a vertical axis wind turbine does not?
    - ❑ **A.** Aerodynamic lift
    - ❑ **B.** Generator
    - ❑ **C.** Gearbox
    - ❑ **D.** Blade pitch control
    - ❑ **E.** None of the above

Rotor Aerodynamics

THE FOCUS OF THIS chapter is the aerodynamics of the rotor. This field of dynamics deals with the movement of air and how it interacts with an object in motion, such as a wind turbine rotor. The basics of two essential wind turbine aerodynamic principles are reviewed, followed by an overview of the aerodynamic features of the rotor. The chapter concludes with a look at basic rotor designs and an examination of the most common of rotor blade problems.

Chapter Topics

This chapter covers the following topics and concepts:

- The basic principles of wind turbine aerodynamics
- The aerodynamic features of the rotor
- Basic rotor designs
- Common blade problems

Chapter Goals

When you complete this chapter, you will be able to:

- Describe the fundamental principles of lift and drag
- Describe the various aerodynamic features of the rotor
- Summarize the characteristics of the two primary rotor blade designs
- Explain typical rotor blade problems

Basic Wind Turbine Aerodynamic Principles

To gain an understanding of how wind turbines function, you need to know that wind turbine technology employs the use of aerodynamics. There are two fundamental aerodynamic principles you must be familiar with in relation to wind turbines: lift and drag.

Lift

Lift is the aerodynamic force that keeps an aircraft in the air. For most wind turbine designs, lift is what ultimately causes the blades to rotate. In wind turbines, the blades are shaped as airfoils, similar to airplane wings. Airfoils have a slightly raised and rounded face and a flatter underside. Because the top surface of the airfoil is rounded, the distance the wind travels across the upper curved surface is a bit longer than the distance over the lower flat surface. This slight difference creates a pressure that results in a force that moves in a perpendicular or upward motion away from the airfoil. This force is called lift because it causes the object to lift or rise. But wind turbine rotors are not like airplane wings. Because the blades are attached to the hub they do not rise. Instead, the lift propels the hub to rotate. Whenever a lift force is created, the aerodynamic force of drag is also present.

Drag

Drag is the aerodynamic force that provides resistance to movement. Unlike lift, which moves perpendicular to wind flow, drag moves in the same direction as the wind. In wind turbines, drag slows down the rotation of the rotor. Ideally, wind turbines should have a high lift and low drag for optimal functioning. Drag also refers to a limited number of small to medium-sized VAWT designs, similar to an anemometer or blower, that use drag exclusively as the means of powering rotation. Some newer designs of VAWTs attempt to use drag and lift at different stages of rotation of each blade in VAWT applications to increase overall efficiency.

Aerodynamic Features of the Rotor

The wind turbine rotor possesses various aerodynamic features, all of which are considered by wind turbine designers. Be aware that when discussing the efficiency of the rotor, the term **rotor power coefficient** (C_{PR}) is used. More specifically, this term describes how well the turbine transforms wind into electricity. The higher the rotor power coefficient, the more efficient the rotor.

Number of Rotor Blades

Three-blade rotors are the most common type of wind turbine today. Why? As you can see in **FIGURE 4-1**, one reason is turbines that employ fewer blades have increased rotational speed.

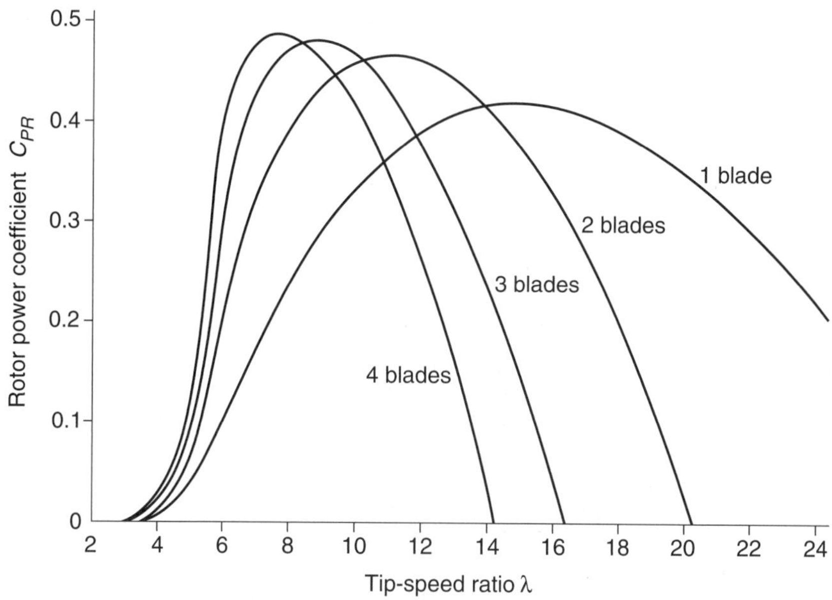

FIGURE 4-1 The increase in the rotor power coefficient consistently decreases with each additional blade.

Adapted from Hau, E. (2006). *Wind Turbines*. Berlin and Heidelberg: Springer-Verlag.

Note in the graph that the tip-speed ratio, represented as the symbol λ, is another factor to consider when assessing the efficiency of a rotor. This concept is explored in more depth shortly. Simply, tip-speed ratio refers to the ratio between the rotational speed of a rotor blade's tip and the speed of the wind. Even though a four-blade rotor functions at a slightly better level than three blades, the additional costs involved in an extra blade would outweigh any benefits in the turbine's functioning.

Three blades are preferable over fewer blades because additional problems accompany one- and two-blade rotors. One or two fewer blades demand extra design considerations. Additionally, one and two blade rotors create substantial noise and thus make their use impractical in most areas. It should be noted that as offshore wind farms increase in popularity, the potential for exploring the use of one- and two-blade turbines exists.

Blade Shape

Researchers have invested much effort in finding the ideal blade shape or geometric form of the blade. Experts refer to the optimum aerodynamic shape as a "hyperbolically delimited trapezoid" as illustrated in **FIGURE 4-2**.

When the ideal shape has been compared against a rectangular blade and a semi-trapezoidal blade, the trapezoid proves to be the most efficient of the three, as shown in **FIGURE 4-3**.

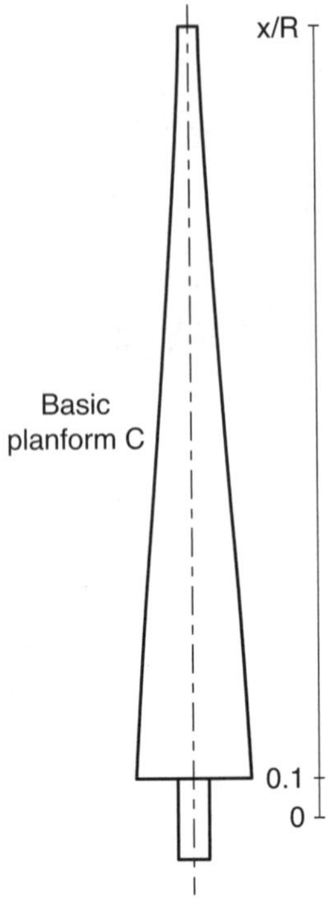

FIGURE 4–2 A blade in the shape of a trapezoid.

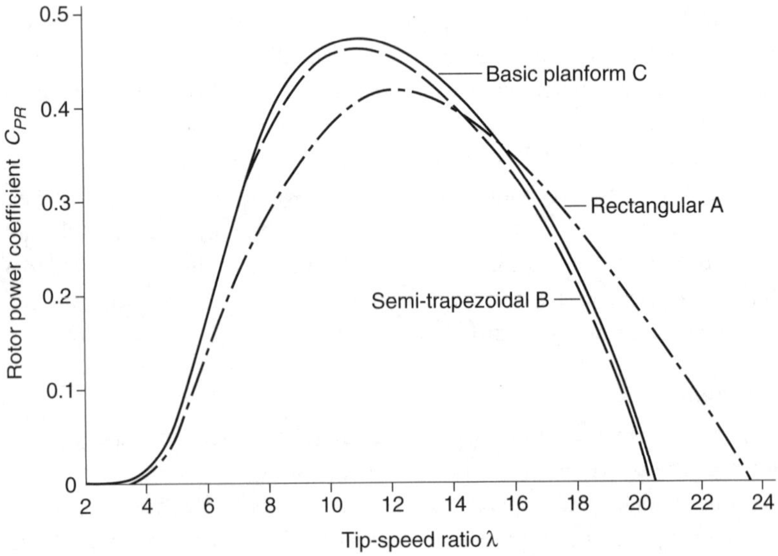

FIGURE 4–3 This figure illustrates how blade shape has a direct effect on the rotor power coefficient.

Adapted from Hau, E. (2006). *Wind Turbines*. Berlin and Heidelberg: Springer-Verlag.

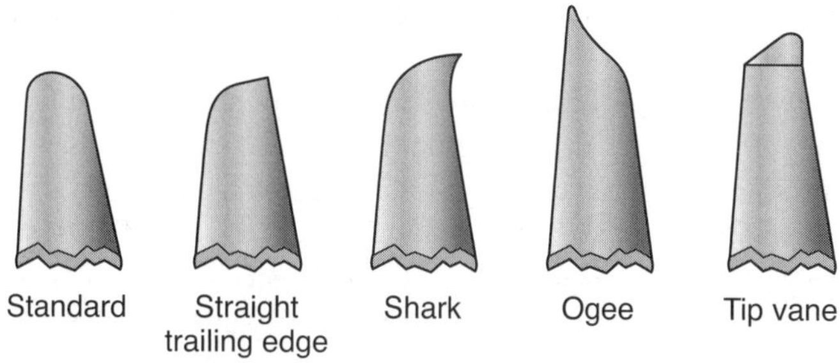

FIGURE 4-4 Examples of various rotor blade tip shapes and a tip vane.
Adapted from Hau, E. (2006). *Wind Turbines*. Berlin and Heidelberg: Springer-Verlag.

Additionally, investigators have discovered that a turbine's power can be improved through the design of the tip. **FIGURE 4-4** contains examples of several different rotor blade tip shapes as well as a tip vane. A similar trapezoidal shape should be applied to the outer blade tip to obtain the best results.

Rotor Blade Twist

Another aerodynamic feature of the rotor is the rotor blade twist. Incorporating a twist into the blade's design has been found to increase the rotor power coefficient. A twist also maximizes wind flow along the whole length of the blade from the blade root to its tip. To calculate the exact angle of the twist involves a detailed examination of a variety of other factors, such as the type of airfoil and operational mechanisms.

One must remember that because wind speed is not fixed, determining the ideal angle of the twist can be calculated only for a particular ratio of peripheral speed to wind speed. This is referred to as the rated power operating point. Under all other different conditions, such as with increases or decreases in wind speed, the twist does not operate at its optimal level, which contributes to losses in power. Variations of the twist exist. Often, the inner area of the blade is not manufactured with a noticeable twist because of the logistics of producing such twists. Others have a substantial twist on the inner area of the blade and very little twist on the outer area. Although it is much simpler to produce a rotor blade without a twist, experts agree that a twist-free blade will contribute to a loss in power.

Rotor Blade Airfoils

When assessing rotor blade airfoils, the variable that carries the most importance is called the lift-to-drag ratio, also written as L/D. Remember that high lift with low drag is optimal. When the L/D ratio decreases, so does the power coefficient. Interesting to note is the difference between low-speed and high-speed rotors. The L/D ratio is of little significance in rotors of lower speeds. However, with high-speed rotors the L/D ratio makes a substantial difference in how much power can be generated.

Until recently, most rotor blade airfoils in use were not designed for wind turbines but rather for aircraft. Although functional similarities exist, they are not identical. It is for this reason that designers today are creating rotor blade airfoils specifically for wind turbines. One important characteristic of airfoils is surface quality. External factors, such as weather, dirt, and insects impact the surface of the airfoil and can cause degradation. As the surface of the airfoil becomes rough, performance declines.

Blade Thickness

Apart from considering the aerodynamic features of a rotor, wind turbine designers must also give equal importance to structural considerations. When looking at aerodynamics and blade thickness, remember that the thinner the blades, the better the rotor functions. But when you take the perspective of structural needs, thickness in the cross section is required for proper support. These conflicting needs make choosing the appropriate thickness a challenge. Design engineers work to optimize the thickness of the blade in order to minimize its overall mass while meeting design requirements for the performance and environmental conditions the turbine is intended for. Many of the same calculations from designing aircraft are used here, modified for the needs of wind turbine blades.

Rotor Tip-Speed Ratio

Recall that tip-speed ratio refers to the ratio between the rotational speed of a rotor blade's tip and wind speed. Today, many wind rotors have a high tip-speed ratio. One reason for this is less **torque** is required with a high rotor tip-speed ratio, which translates to a physically lighter rotor shaft and gearbox. Another advantage of high tip-speed ratio is cost savings. Overall, fewer materials for constructing the rotor are required when the tip-speed ratio is high. However, these savings are lost with the necessity to employ advanced and costly blades. Standard blades are not suitable for rotors with a high tip-speed ratio. Interestingly the rotor power coefficient basically remains the same despite variations in tip-speed ratio (see FIGURE 4-5.)

One more factor involved with tip-speed ratio is aerodynamic noise emission. As the tip-speed ratio increases, so does aerodynamic noise emission, which is an important factor to consider when choosing a suitable location for a turbine.

Rotor Blade Design

Rotor blades vary from turbine to turbine. Their design is the result of weighing many factors, such as the exact geometric shape, blade material, and costs involved. Generally speaking, most rotor blades are trapezoidal. As far as typical blade

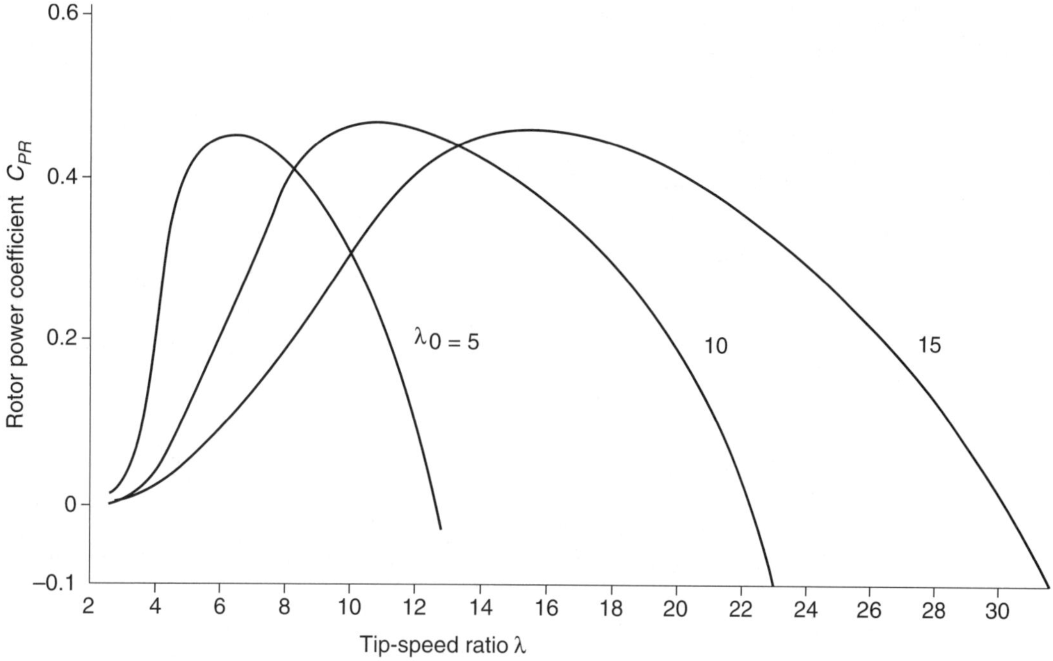

FIGURE 4-5 Even as the tip-speed ratio increases, the rotor power coefficient remains relatively stable.
Adapted from Hau, E. (2006). *Wind Turbines.* Berlin and Heidelberg: Springer-Verlag.

thickness goes, the outer section of the rotor blades has an approximate 12 to 15 percent thickness ratio; the inner section of the blades, near the base of the blade, is usually thicker than the outer section. With regard to the taper of the rotor blade there is an aerodynamic optimum but some manufacturers prefer a taper that is less severe. Another area of the rotor blade that has been receiving more attention is the shape of the blade tip. Aerodynamically optimized blade tips, usually trapezoidal in shape, are standard in today's rotors as they reduce aerodynamic noise.

Three-Bladed Rotors

Certain older Danish wind turbines, such as the Aerostar, have a low tip-speed ratio as well as a low **aspect ratio**, a term that refers to blade slenderness. In newer and larger three-bladed rotors, the aspect ratio is much higher. Generally, these types of rotors have an ideal tip-speed ratio of seven. Blade tips on three-bladed rotors control noise emission up to 70 m/s.

Two-Bladed Rotors

In the 1980s, when two-bladed rotor use was more widespread, different geometric shapes were tested, instead of today's relatively standard trapezoid. The amount of taper varied quite a bit as well.

COMMON BLADE PROBLEMS

Can you imagine a 150-foot blade from a towering wind turbine breaking right off its hub? It sounds like the makings of a horror film, but this didn't happen on a movie set. In March 2010 in Scotland, at one of the largest wind farms in Europe, Whitelee Wind Farm (see **FIGURE 4-6**), a 14-ton blade from one of its 140 turbines fell to the ground. Investigators determined that a manufacturing defect was the culprit. More specifically, it appears that the epoxy resin used on the outer layer of the blades to provide a hard laminate covering was not properly applied during manufacturing. The improper application resulted in a weakness in the blade that led to its fracture.

Blade problems resulting from manufacturing defects, such as hairline cracks or wrinkles in the blade, are commonplace. Damage can occur as well during transport or construction. Such problems increase the cost of wind energy for consumers. Investigation is ongoing to minimize repairs and shutdowns of wind farms due to manufacturing defects and damage.

In addition to manmade errors, there are blade problems to which Mother Nature contributes. For larger wind turbines, avoiding lightning strikes is practically impossible. When lightning strikes a blade, it usually hits the blade tip and causes extensive damage. For this reason, the latest rotor blades come equipped with a lightning protection system that is incorporated into the blade tip. This system contains a lightning conductor that ultimately grounds the system.

FIGURE 4-6 Whitelee Wind Farm in Scotland, where a manufacturing defect led to a severed blade.

COMMON BLADE PROBLEMS (Continued)

Another common blade problem is erosion that occurs from prolonged exposure to the elements, such as rain, hail, and sand. This type of blade damage contributes to turbulence, which reduces the optimal functioning of the turbine. Sometimes, erosion can occur as early as two years after installation. Wind turbine maintenance specialists are currently investigating ways to prevent erosion and thus loss of power output.

CHAPTER SUMMARY

Rotor aerodynamics for wind turbines follow various principles, but the two most basic are lift and drag. Remember that whenever there is lift, drag also exists. Lift is the force that causes rotation of the rotor and drag is the force that impedes it. Optimally, a wind turbine should have high lift and low drag.

A look at the aerodynamic features of the rotor revealed that three-bladed rotors are the most common today; the ideal geometric shape of the blade is trapezoidal; blade twist increases efficiency; rotor tip-speed ratio in most large turbines today is high; and thin blades are ideal from an aerodynamic standpoint but impractical structurally. Rotor blade design varies among turbines with the most striking differences found between three-bladed and two-bladed rotors. Common blade problems include erosion, lightning, and manufacturing defects.

KEY CONCEPTS AND TERMS

Aspect ratio
Drag
Lift
Rotor power coefficient
Torque

CHAPTER ASSESSMENT: ROTOR AERODYNAMICS

1. The newest rotor blades are immune against the external elements that lead to degradation of the surface.
   - ❑ **A.** True
   - ❑ **B.** False
2. The most important parameter to consider when assessing rotor blade airfoils is:
   - ❑ **A.** L/D.
   - ❑ **B.** D/L.
   - ❑ **C.** D x L.
   - ❑ **D.** None of the above
3. What is the ideal lift-to-drag ratio?
   - ❑ **A.** High lift, high drag
   - ❑ **B.** High lift, low drag
   - ❑ **C.** Low lift, low drag
   - ❑ **D.** Low lift, high drag
4. What is the optimal shape of a rotor blade?
   - ❑ **A.** Asymmetrical rectangle
   - ❑ **B.** Trapezoid
   - ❑ **C.** Rectangle
   - ❑ **D.** Pointed trapezoid

5. What are the typical problems with blades? (Select three.)
   - ❑ A. Erosion
   - ❑ B. Collision
   - ❑ C. Overheating
   - ❑ D. Manufacturing defects
   - ❑ E. Lightning
6. The term refers to blade slenderness.
   - ❑ A. aspect ratio
   - ❑ B. tip-speed ratio
   - ❑ C. drag
   - ❑ D. lift
7. Blades are at thickest at the base.
   - ❑ A. True
   - ❑ B. False
8. Which of the following is the most common type(s) of rotor today?
   - ❑ A. One-bladed rotors
   - ❑ B. One- and two-bladed rotors
   - ❑ C. Two- and three-bladed rotors
   - ❑ D. Three-bladed rotors
   - ❑ E. None of the above
9. Drag moves perpendicular to wind flow.
   - ❑ A. True
   - ❑ B. False
10. The aerodynamic design of the rotor blade's reduces noise emission.
   - ❑ A. hub
   - ❑ B. tip
   - ❑ C. leading edge
   - ❑ D. thickness

Wind Resource

chapter 5

WIND SPECIALISTS OVER THE last 20 years have investigated the various factors that affect using the wind as a resource. Previously, wind data came exclusively from meteorological sources. As you will see in this chapter, the weather is not the only factor that affects wind, especially as it relates to wind turbines. This chapter begins with an exploration of the most important characteristics of wind resource followed by an overview of the atmospheric boundary layer. The chapter delves into the various factors that influence wind resource, including annual and seasonal variations, turbulence, geographical variations, and terrain.

Chapter Topics

This chapter covers the following topics and concepts:

- The primary characteristics of wind resource
- The atmospheric boundary layer
- How annual and seasonal patterns cause wind variations
- What turbulence is and how it affects wind resource
- How geographical variations affect wind resource
- How terrain affects wind resource

Chapter Goals

When you complete this chapter, you will be able to:

- Explain the primary characteristics of wind resource
- Define and describe the atmospheric boundary layer
- Outline the annual and seasonal wind variations
- Explain the phenomena of turbulence
- Describe the various geographical variations in wind resource
- Explain the effect of terrain on wind resource

Characteristics of Wind Resource

In order to use wind as an alternative source of energy, understanding its characteristics is essential. Without question, the most important trait of wind resource is its variability. The wind is extremely changeable under various circumstances and settings. Having an awareness of these situations is important. You can think of variability in terms of the following two broad categories:

- Spatial variability—This type of variability deals with geography. The world consists of a variety of different climates that affect how windy a particular region is. The specific climate in a region is also variable thanks to topography or the physical features of specific area. Wind is particularly sensitive to topography. For instance, more wind exists at greater heights, such as at the top of a mountain, and obstacles such as urban structures reduce wind speed. You will read more about how terrain affects wind further along in this chapter.
- Temporal variability—This type of variability addresses the changes in wind over time. Consider that in the same location, wind patterns can be different from year to year or even over longer periods of time, such as 10, 15, or 20 years. At present, wind energy specialists know little about these long-term variations. This limits effective long-term planning for wind farms. Specialists know a bit more about seasonal variations, which will be explained further shortly. However, making accurate predictions more than a few days ahead (known as **synoptic variations**) is challenging due to the movement of weather systems such as storms. Wind changes throughout the day, also called **diurnal variations**, occur in some locations and can often be predicted. Changes in wind speed can also occur within minutes, seconds, or even less than seconds. These variations are known as **turbulence**, which is explored in more depth later in the chapter.

The Atmospheric Boundary Layer

The atmospheric boundary layer, often just referred to as the boundary layer, is important in understanding turbulence. You can think of the boundary layer as the slice of atmosphere that sits between the ground and the layer that sits above all obstacles, urban or natural. The wind that flows in this obstacle-free layer is called **geostrophic wind**. Below this airflow, closer to the earth, is where you find the boundary layer, which ranges between 600 and 2,000 meters. It is in this layer that the wind interacts with the earth.

A subsection of the boundary layer, known as the **Prandtl layer**, is the area that sits close to the earth's surface. The height of the Prandtl layer is not fixed. During the day it ranges between 50 and 150 meters but in the evening it drops to 10 to 50 meters.

Velocity

When examining wind speed or velocity within the boundary layer, note that velocity increases with elevation. However, when examining increases in wind speed, other factors must be looked at in conjunction with elevation. An important factor to consider is surface roughness, which is expressed by roughness length or z_0. Surface roughness is also categorized into classes, beginning at 0, the lowest possible. A higher roughness length (and roughness class) causes wind speeds to decrease. TABLE 5-1 shows the different roughness lengths and roughness classes for several different type of terrain.

Temperature

In the boundary layer, temperature behaves in different ways and is grouped into three categories:

- **Unstable stratification**—It begins with an increase in surface heat. This makes warm air near the ground rise. With increased height, pressure causes the air to rise and thus, cool. However, if the natural cooling process isn't enough to balance the temperature, the air continues to rise. The rising air then produces large cells, which contribute to significant turbulence in the boundary layer.
- **Stable stratification**—It begins the same way as unstable stratification except that the natural cooling process that occurs with rising heat is sufficient to cool the air so that it is colder than the environment. This normally occurs in the evenings.
- **Neutral stratification**—It occurs in the same fashion as the previous two types of stratification except that when the warm air rises, it cools to the same temperature as the surroundings. Typically, this occurs in the presence of strong wind. Turbulence results from surface roughness combined with the boundary level.

TABLE 5-1 ROUGHNESS LENGTHS AND ROUGHNESS CLASSES FOR DIFFERENT TYPES OF SURFACES

z_0 [m]	Types of terrain surfaces	Roughness class
1.00	City	> 3
0.50	Suburbs	3
0.30	Built-up terrain	3
0.20	Many trees and/or bushes	> 2 and < 3
0.10	Agricultural terrain with a closed appearance	2
0.05	Agricultural terrain with an open appearance	1
0.03	Agricultural terrain with very few buildings, trees, etc.	1
0.01	Airports, runway	<1
$5 * 10^{-3}$	Bare earth (smooth)	<1
10^{-3}	Snow surfaces (smooth growth)	<1
$3 * 10^{-4}$	Sand surfaces (smooth)	0
10^{-4}	Water surfaces (lakes, fjords, and the sea)	0

Data from Hau, E. (2006). *Wind Turbines*. Berlin and Heidelberg: Springer-Verlag.

Each is important to consider when assessing wind energy. Neutral stratification is of the most importance because turbulence is at its greatest in strong winds. This is not to say that the other two categories are not important. Unstable stratification can cause unexpected gusts of wind. Stable stratification can lead to unbalanced wind loads on a turbine.

Relative Humidity

Humidity, or the amount of moisture in the air, is another variable factor within the boundary layer.

Annual and Seasonal Variations

As mentioned earlier, wind speeds vary from year to year and season to season. At present, the variability that occurs from year to year cannot be accurately predicted. Seasonal variation is not fixed throughout the world; wind speed is specific to its geographic location. For medium continental latitudes, wind speed is at its highest from about December through March. Predictions of changes in wind speed throughout the year are generally accurate using statistics to calculate probability.

Wind Speed Predictions

Wind speed predictions can be classified into two general groups:

- **Short-term predictions**—Forecasts that relate to seconds or minutes in the future. Such predictions can be helpful in relation to the operational functioning of wind turbines or wind farms. These forecasts are determined through the use of statistical techniques that involve looking at trends in the recent past.
- **Long-term predictions**—Forecasts that relate to hours or days. These predictions can come in handy in the planning stages of expanding a power station network. Although statistical measures are used to make long-term predictions, meteorological methods are also used. Combining both of these techniques can often provide accurate information of the power output of a wind turbine or farm.

Long-Term Wind Speed Variations

Extensive analysis has shown that wind speed, regardless of location, goes through slow, long-term variations. Investigators have uncovered certain trends that are linked to long-term temperature changes. However, current predictions maintain that global climate change will influence wind patterns in the near future. Other natural occurrences, such as volcanic eruptions, also alter the atmosphere, which ultimately affects wind. Such changes make it even more difficult to accurately calculate the energy output of a wind farm over an extended period of time.

Turbulence

Understanding how turbulence plays a role in wind resource is essential, as turbulence can strain wind turbine operation. As mentioned earlier, turbulence is defined as variations in wind speed that occur within minutes, seconds, or even less than seconds. Generally, the time frame for such fluctuations is less than 10 minutes. Turbulence is generated by the wind's interaction with topography and the influence of temperature. Frequently, the two causes of turbulence occur at the same time.

Turbulence Intensity

Measuring turbulence is done statistically. Turbulence intensity refers to a general amount of turbulence at one time. The formula for turbulence intensity or I is:

$$I = \frac{\sigma}{U}$$

Turbulence intensity = standard deviation of wind speed variations/mean wind speed

Note that the mean wind speed is taken over 10 minutes or 1 hour. Remember that this simple formula cannot be used to determine turbulence intensity under all circumstances. This formula provides you with overall level of turbulence. Keep in mind that surface roughness, height above ground, terrain, and temperature all interact and can affect turbulence intensity.

Geographical Variations in Wind Resource

Recall that solar energy and its heating of the earth is what generates wind. The sun's heat is not equally distributed over the earth's surface; thus wind resource will not be equal geographically. The highest temperatures are found on land near the equator during the day. The warm air from this area rises, mixes with the atmosphere, and ultimately cools other geographical areas. Additionally, the rotation of the earth assists in forming a global circulation pattern that contributes to differences in climate.

On a smaller scale, elevated land, such as hills and mountains, tend to have increased wind speeds. In contrast, valleys usually contribute to decreased wind speeds. Coastal regions typically show increased wind speeds. Local circulation patterns are created along coastlines because the land and ocean are not the same temperature.

Wind Maps

Wind maps, or wind atlases, are one tool used to assess whether a particular geographic area would be conducive to a wind turbine or farm. In order to read a wind map, you need to know that North American wind maps commonly use wind power classes to describe the potential of the wind for different geographical areas. A common system in use today is the Battelle classes of wind power density, shown in **TABLE 5-2**. Classes are determined by wind power per square meter of the wind stream (W/m^2) at various heights.

As the table indicates, wind classes range from 1 (the lowest) to 7 (the highest). The minimum wind class required for wind farms is Wind Class 3.

European wind specialists use their own European Wind Atlas instead of the North American Battelle class system. They use a color-coded system that consists of five zones, five classes of surface roughness, and the mean annual wind speed at 50 meters high. Note that the European Wind Atlas does not include Eastern Europe.

Wind Farm Locations

Wind farms are located throughout the world, but as you know by now, they cannot be built just anywhere. Before looking at various wind farm locations, review the areas that are ideal for "harvesting" energy from the wind.

TABLE 5-2 THE BATTELLE CLASSES OF WIND POWER DENSITY

	Wind speed and power at 10 m (33 ft)			Wind speed and power at 30 m (100 ft)			Wind speed and power at 50 m (164 ft)		
Wind power class	Power density	Speed		Power density	Speed		Power density	Speed	
	W/m²	m/s	mph	W/m²	m/s	mph	W/m²	m/s	mph
	50	3.5	7.8	80	4.1	9.2	100	4.4	9.9
1									
	100	4.4	9.9	160	5.1	11.4	200	5.5	12.3
2									
	150	5.0	11.5	240	5.9	13.2	300	6.3	14.1
3									
	200	5.5	12.3	320	6.5	14.6	400	7.0	15.7
4									
	250	6.0	13.4	400	7.0	15.7	500	7.5	16.8
5									
	300	6.3	14.1	480	7.4	16.6	600	8.0	17.9
6									
	400	7.0	15.7	640	8.2	18.4	800	8.8	19.7
7									
	1,000	9.5	21.3	1,600	11.1	24.7	2,000	11.9	26.7

Note: Increase in speed and power with height assumes $1/7$ power law. Wind speeds are based on the equivalent of a Rayleigh distribution.
Data from Gipe, P. (2009). *Wind Energy Basics.* White River Junction, VT: Chelsea Green Publishing Company.

Areas Conducive for "Harvesting" Energy from the Wind

When examining ideal areas for wind power technology, you obviously need an area that experiences regular wind. But that is not the only factor. Wind speeds must be examined as well. Good sites for wind farms are where winds have the opportunity to rise to a higher altitude. Examples include: along a coastline, an open field, along a hill, or in a gap between mountains. Wind can be harvested if the area is flat or rounded and without obstacles.

America

As shown in **FIGURE 5-1**, the United States has several areas that are conducive to wind farms. The West Coast boasts excellent conditions. Successful wind farms are located in California, Oregon, and Washington, where wind classes are 3 and higher. In fact, a town in California called Inyokern is considered to be the best location for a wind farm in the entire country. The East Coast from North Carolina up to Maine has ideal conditions with wind classes at 4 and higher. In the midsection of the country is the Great Plains, which provides a good spot for wind farms. States in this region and the surrounding areas include Kansas, Wyoming, Nebraska, Iowa, Minnesota, and Texas. These areas are usually a class 4 or 5. The areas around the Great Lakes have wind classes 4 and higher. Some areas in Alaska, particularly over the Aleutian Islands, show a wind class as high as 7. Parts of Hawaii are classified as a 6. In Puerto Rico, the Virgin Islands, and the Pacific Islands wind classes are generally 3 or 4.

Europe

If you take a look at the European Wind Atlas, you see that the northernmost countries—Finland, Sweden, Germany, Denmark, and Britain—possess the best conditions for wind farms. A closer look reveals that certain parts of Spain, the south of France, and Greece are also ideal for harvesting wind. With many ideal locations already in use in Western Europe, wind energy projects are being explored in Eastern Europe in countries such as Bulgaria and Hungary, and in Turkey.

Australia

Australia's greatest wind resource lies in the southern part of the country, in the southwest of Western Australia and southern South Australia. South Australia is home to more than half of the country's wind resource. Other areas that have strong wind speeds include western Victoria, northern Tasmania, and high-altitude areas of New South Wales and Queensland. Harvesting wind resource in Victoria is currently expanding.

Asia

Within Asia, China leads the way in wind farms. China is the second-largest producer of wind energy; the United States is the first. China's wind resources are plentiful because of its extensive coastline and large land mass. India, which also has an extensive coastline, is the fifth-largest producer of wind energy.

Russia

Strong winds are found in the northwest region, along the southeastern coastlines, and in parts of the southwestern coastline that borders the Black Sea.

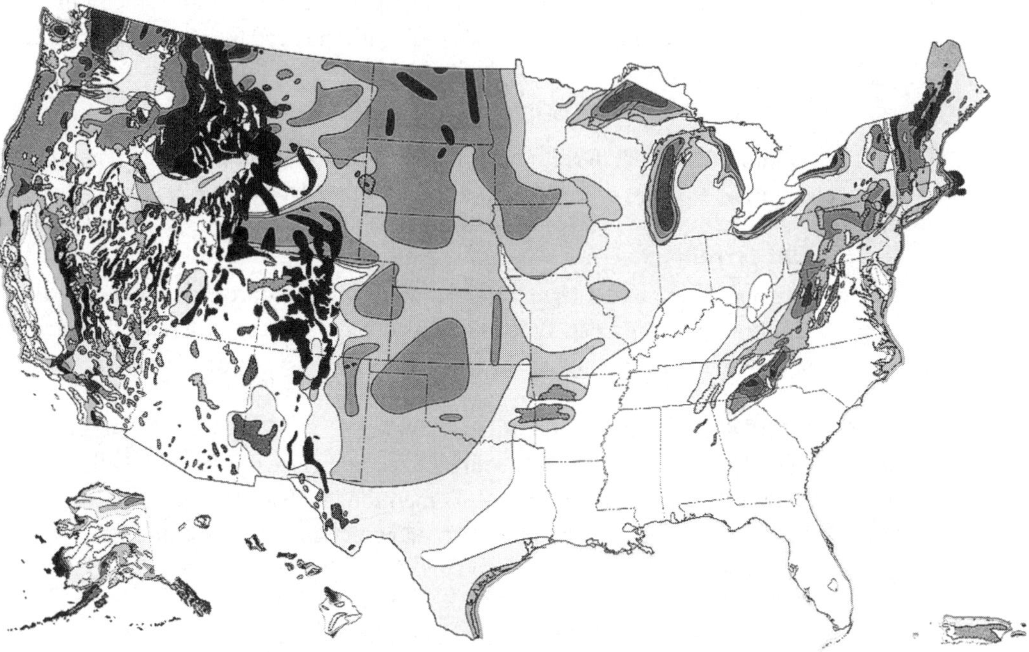

10 m (33 ft)

Wind Power Class	Wind Power Density (W/m²)	Speed(b) m/s (mph)
	0	0
1	100	4.4 (9.8)
2	150	5.1 (11.5)
3	200	5.6 (12.5)
4	250	6.0 (13.4)
5	300	6.4 (14.3)
6	400	7.0 (15.7)
7	1000	9.4 (21.1)

FIGURE 5-1 A wind map of the United States illustrates the annual average wind power.
Courtesy of NREL

Israel

The unique geology of the Kinneret (Sea of Galilee) and the mountains to the east lead to ideal conditions for wind power. As winds from the Mediterranean and northern Israel drop as they enter what is the northern end of the Great Rift Valley, with the Kinneret more than 200 meters below sea level, the air compresses to unusually high density and heats. As the sun heats the air during the day, air is forced up and over the mountains consistently and with best velocity in the late afternoon, when local power needs peak with the need for air conditioning.

Effect of Terrain on Wind Resource

As indicated earlier, terrain has a marked effect on wind resource. The following examines different types of terrain and how they affect wind resource.

Flat Terrain

When speaking of flat terrain in relation to wind resource, recognize that the terrain is not perfectly flat. Terrain is considered flat if any differences in the height of the ground are no higher than 60 meters within a radius of 11.5 km.

Changes in Surface

Changes in the ground's surface are related to surface roughness and the existence of objects, natural or manmade. As mentioned, obstacles such as trees or buildings produce severe turbulence that can potentially damage a turbine.

Hilly Terrain

Ground that is especially hilly can contribute to increased turbulence.

GREEN ENERGY POWERS COASTAL COMMUNITY SERVICES

Should wind technology become a required area of study for social workers? If what's happening in Washington State is any indication, wind power could prove to be of great benefit to the social service realm and the communities that access it. Starting on June 1, 2010, in Grayland, the Coastal Community Action Program, a social service agency, played a key role in getting four large wind turbines to generate power for around 700 homes. The turbines were built with the intention of providing assistance to low-income families under an initiative called the Coastal Energy Project, the first of its kind in the country.

Power from these turbines is channeled into the local Grays Harbor PUD grid and the money that is subsequently earned is designated for social services programs, such as Meals on Wheels, Energy Assistance, and other social programs for people in financial need. The Coastal Energy Project receives about $500,000 for powering the local community. The project received a one-time $5 million grant, in addition to funding from other energy programs, as well as investments. Wind energy could be just the avenue to finally meet the needs of neglected social service agencies on a long-term basis.

Chapter Summary

Accessing wind resource involves a detailed understanding of the conditions that influence it. Of great importance is the variability of wind resource, both spatially and temporally. A review of the boundary layer and how it interacts with velocity and temperature provides insight into the phenomena of turbulence. The challenges to planning include seasonal and year-to-year wind variations. A look at the variations in geography throughout the world illustrates the importance of wind maps and their role in accessing wind resource. Remember that terrain plays an integral role in wind velocity and thus the amount of wind resource available.

Key Concepts and Terms

Diurnal variations
Geostrophic wind
Prandtl layer
Synoptic variations
Turbulence

Chapter Assessment: Wind Resource

1. Turbulence is generated by the wind's interaction with topography and the influence of temperature.
- ❑ **A.** True
- ❑ **B.** False

2. For medium continental latitudes, when is wind speed is at its highest?
- ❑ **A.** June through to September
- ❑ **B.** September through to December
- ❑ **C.** December through to March
- ❑ **D.** March through to June

3. Changes in wind speed that occur throughout a day are called _______ variations.
- ❑ **A.** geostrophic
- ❑ **B.** synoptic
- ❑ **C.** turbulent
- ❑ **D.** diurnal

4. The Prandtl layer is the slice of atmosphere that sits above all obstacles.
- ❑ **A.** True
- ❑ **B.** False

5. What are two broad categories of variability? (Select two.)
   - ❑ A. Temporal
   - ❑ B. Meteorological
   - ❑ C. Visual
   - ❑ D. Spatial
6. A town in _______ is thought to have the absolute best conditions for accessing wind resource.
   - ❑ A. Alaska
   - ❑ B. Maine
   - ❑ C. Texas
   - ❑ D. California
7. What conditions or areas are conducive for harvesting wind? (Select three.)
   - ❑ A. Along a coastline
   - ❑ B. A plain
   - ❑ C. Building rooftop
   - ❑ D. In a gap between mountains
   - ❑ E. On the periphery of a wooded area
8. China is the second largest producer of wind energy today.
   - ❑ A. True
   - ❑ B. False
9. The height of the boundary layer sits between 600 and _______ meters.
   - ❑ A. 1,200
   - ❑ B. 1,500
   - ❑ C. 2,000
   - ❑ D. 3,000
10. Under the Battelle wind class system, what is the minimum class needed to access wind resource?
    - ❑ A. 2
    - ❑ B. 3
    - ❑ C. 4
    - ❑ D. 5
11. What factor does the European Wind Atlas include that the North American Battelle class system does not?
    - ❑ A. Temperature ranges
    - ❑ B. Classes of surface roughness
    - ❑ C. Classes of humidity
    - ❑ D. Turbulence markers
12. Coastal regions usually show increased wind speeds because of circulation patterns caused by differential temperatures between the land and the ocean.
    - ❑ A. True
    - ❑ B. False

13. For terrain to be considered "flat," what is the maximum difference in height within a radius of 11.5 km?

- ❑ **A.** 60 meters
- ❑ **B.** 50 meters
- ❑ **C.** 40 meters
- ❑ **D.** 30 meters

14. At present, the variability that occurs from year to year can be predicted with relative accuracy.

- ❑ **A.** True
- ❑ **B.** False

15. Within the boundary layer, when rising heat is sufficient to cool the air so that it is colder than the environment, this is known as _______ stratification.

- ❑ **A.** neutral
- ❑ **B.** unstable
- ❑ **C.** stable
- ❑ **D.** None of the above

chapter 6

Power Output and Energy Yield

THE SUCCESS OF A wind farm depends on a variety of factors, but the focus of this chapter deals with just one: how to estimate the performance of a wind turbine. More specifically, how to estimate the gross amount of energy that wind turbines can make use of. After an overview of the three broad categories of wind speed, you will read about the three most common methods for estimating performance. It should be noted that a performance estimate is just that—an estimate. With wind being such a variable resource, an estimate in lieu of a prediction is the best specialists have been able to achieve to date. Finally, remember the difference between power and energy. Power is a measure of energy in the moment (watts) and energy describes the accumulation of power over a specified time period (watt-hours).

Chapter Topics

This chapter covers the following topics and concepts:

- Key terminology related to wind speed
- The ways you can estimate wind turbine performance

Chapter Goals

When you complete this chapter, you will be able to:

- Define key terminology related to wind speed
- Identify the three primary methods of estimating wind turbine performance
- Explain what constitutes the swept area method
- Describe the limitations in using the power curve method
- Outline how to estimate wind performance using manufacturer's estimates for both small and large wind turbines

Key Terminology

Recognizing the characteristics of wind according to three broad categories gives you more knowledge with which to estimate performance. When discussing wind speed, the three types are instantaneous, average, and peak.

Wind Speed

Wind speed or **instantaneous wind speed** is the velocity of the wind in the moment. This type of wind speed focuses solely on the present. Wind speed from five hours, five minutes, or even five seconds ago is not being measured. Averages are not used. Because wind is not fixed and can vary wildly, even second to second, instantaneous wind speed cannot be used to estimate the amount of wind resource available in the wind. Instantaneous wind speed when absent sits at zero. Regular winds, which are typical of most sites, range between 5 and 35 mph. Winds above this range could be related to storms and can go up to 100 mph or higher. Those speeds sound impressive but this information on its own offers no help in estimating performance.

Average Wind Speed

In contrast to instantaneous wind speed, **average wind speed** provides the type of information necessary to predict how much energy a wind turbine will produce. Simply, an average of instantaneous wind speeds are taken within a specific time frame. In a simple example, say you measure the wind speed at ten different times over an hour. The results give you one third of the measurements at 10 mph, another third at 15 mph, and the final third measures in at 20 mph. The math looks like this:

Average wind speed = (10 + 15 + 20)/3
Average wind speed = Individual measurements of instantaneous wind speeds/ the number of measurements taken

Average wind speed for smaller wind turbines or home-scale turbines is a useful tool for estimating the power output of a wind generator. But average wind speed is not an adequate measure for larger turbines and wind farms. The average wind speed for locations of home-scale wind generators ranges from 6 mph to 15 mph.

Peak Wind Speed

The last category of wind speed types is **peak wind speed**. Just as its name implies, it calculates the highest wind speed that could potentially occur in a particular location. This information is crucial for the survival of the turbine. You need to know the highest winds a turbine must be able to withstand. Peak wind speed is also useful for making the most of strong winds. Remember that wind speed is "cubed," so to speak; gusts can greatly intensify the power in the wind. Knowing the peak wind speed helps turbine owners or operators maximize ideal conditions to garner power from the wind.

Estimating Performance

As mentioned in the introduction to this chapter, estimating performance is not the same as making a prediction. It has been found that small wind turbines, particularly those that use a battery-charging system, utilize approximately 70 percent of available energy. Fully charged batteries cannot take on available energy from the wind. When this happens, the energy is often spilled back out into the wind. Some small wind turbines, such as the micro and the mini, include the feature of a dump load, which is a circuit where the excess energy is stored until needed.

There are three primary methods of estimating how a wind turbine will perform: the swept area method, the power curve method, and the use of estimates from manufacturers.

Swept Area Method

To properly apply the first method of estimation, the swept area method, you need to know the power equation. To find power in the wind you use the following equation:

$$P = \tfrac{1}{2}\,\rho A V^3$$

Power in the wind = ½ × air density × intercept area/swept area × speed cubed

To find the swept area you follow this equation:

$$A = \pi r^2$$

A = 3.14 (pi) × circle's radius (half the circle's diameter) × circle's radius

In order to estimate how much energy is available in the wind over a particular time period you use this equation:

$$E \text{ (Energy)} = Pt$$

Energy (kWh/m^2) = Average power (W/m^2) × Time (hours)

Time is represented in hours, so to determine the annual energy of a turbine, time equals 8,760 hours:

$$\text{Energy} = \tfrac{1}{2}\,\rho A V^3 \times 8{,}760$$

Small wind turbines are usually situated in locations that average wind speeds of 9 to 11 mph (4.0 to 5.0 m/s). Say that you want to estimate the amount of electricity a small wind turbine might produce over a year. Its average wind speed is 5 m/s, which corresponds to a power level of 146 W/m² (see **TABLE 6-1**). Note that energy is measured in kilowatt-hours so you must divide power, currently in watts, by 1,000. The equation is as follows:

$$E = .146 \times 8{,}760$$
$$E = 1{,}279 \text{ kWh/year}$$

This quick calculation gives you the estimate that this hypothetical small wind turbine's rotor would intercept an average of approximately 1,279 kWh over a period of a year. Be careful not to confuse how much energy is in the wind stream to actual energy output (AEO) or annual energy production (AEP). The following equation calculates the approximate AEO for a small wind turbine that has an average annual wind speed of 5.0 m/s:

$$\text{AEO} = \tfrac{1}{2}\,\rho A V^3 \times 8{,}760 \times 1.9 \text{ (for Rayleigh distribution factor)} \times 20\% \text{ (conversion efficiency)}/1{,}000 \text{ (from watts to kilowatts)}$$
$$\text{AEO} = 146 \times 8{,}760 \times 1.9 \times 20\%/1{,}000$$
$$\text{AEO} = 260 \text{ kWh/year}$$

TABLE 6-2 and **TABLE 6-3** outline the calculations for small and large wind turbines, respectively:

TABLE 6-1 ANNUAL ENERGY DENSITY BASED ON THE RAYLEIGH PROBABILITY DISTRIBUTION

Annual average wind speed (m/s)	(mph)	Annual power density (W/m²)	Annual energy density (kWh/m²)
4	9.0	75	656
5	11.2	146	1,281
6	13.4	253	2,214
7	15.7	401	3,515
8	17.9	599	5,247
9	20.2	853	7,471

Data from Gipe, P. (2004). *Wind Power: Renewable Energy for Home, Farm, and Business.* White River Junction, VT: Chelsea Green Publishing.

TABLE 6-2 ESTIMATED AEO FOR SMALL WIND TURBINES

Average annual wind speed		Power density	Total efficiency	Average annual specific yield
m/s	~mph	W/m²	%	kWh/m²/yr
4.0	9.0	75	19	120
5.0	11.2	146	20	260
6.0	13.4	253	18	400
7.0	15.7	401	17	600
8.0	17.9	599	15	790
9.0	20.2	853	14	1,050

Note: Gross generation for a single turbine at hub-height wind speed, based on published manufacturer's data. Actual performance will vary.

Data from Gipe, P. (2009). *Wind Energy Basics.* White River Junction, VT: Chelsea Green Publishing Company.

Power Curve Method

The second method of estimating performance is the power curve method. When a wind turbine's power curve is depicted on a graph, as shown in **FIGURE 6-1**, you see a curve that reaches a peak and then drops.

TABLE 6-3 ESTIMATED AEO FOR LARGE WIND TURBINES

Average annual wind speed		Power density	Total efficiency	Average annual specific yield
m/s	~mph	W/m²	%	kWh/m²/yr
4.0	9.0	75	35	230
5.0	11.2	146	37	470
6.0	13.4	253	35	770
7.0	15.7	401	31	1,090
8.0	17.9	599	26	1,360
9.0	20.2	853	21	1,570

Note: Gross generation for a single turbine at hub-height wind speed, based on published manufacturer's data. Actual performance will vary.

Data from Gipe, P. (2009). *Wind Energy Basics.* White River Junction, VT: Chelsea Green Publishing Company.

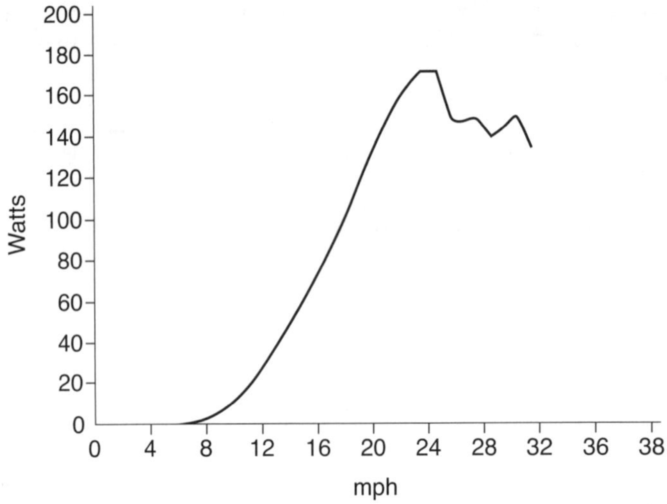

FIGURE 6-1 A power curve that illustrates estimated performance for the Southwest Windpower's Air Breeze micro wind turbine.

Adapted from Gipe, P. (2009). *Wind Energy Basics.* White River Junction, VT: Chelsea Green Publishing Company.

Commonly, however, you see the power curve not represented as a curve on a graph but presented as a table (see **TABLE 6-4**).

Both depictions show the most important piece of information: the wind speed that results in the peak power.

Why the Power Curve Can Be Misleading

The power curve method is far from perfect. Although it may act as an estimate, the power curve can also be misleading. No regulatory agency checks the accuracy of a manufacturer's stated power curve. Keep this in mind when examining power curve graphs or tables from a turbine manufacturer. An independent laboratory, ideally one associated with an educational institution, should validate the information. If a third party hasn't verified the data found on a power curve, be wary of accepting it. Aside from the lack of regulation, there are three other areas that can make the power curve method problematic.

Rated Power **Rated power** is a term commonly used in the wind turbine industry and it usually reflects the peak watts in a power curve. The power peak in a curve is not the same as energy output, although it can be presented in such a way to seem synonymous. In fact, when tested in the field, many small wind turbines do not even demonstrate the rated power as advertised by the manufacturer. However, note that the rated power figures for most large wind turbines do reach their rated power figures. What's important to remember is that rated power does not equal energy production.

TABLE 6-4 THE ANNUAL ENERGY PRODUCTION FOR THE MICRO TURBINE AIR BREEZE

	Rotor diameter		Area		Average wind speed	
	m	ft	m^2		m/s	mph
	1.17	3.83	1.07		5.4	12
Wind speed (mph)	**Measured power (W) 15°C at sea level**	**Raleigh frequency of occurrence**	**Hours/yr**	**Energy (kWh/yr)**	**Gross energy (kWh/yr)**	**System efficiency**
5	0.1	0.048	417	0	3	0.02
10	15.7	0.063	554	9	33	0.27
15	67.8	0.048	420	28	83	0.34
20	144.8	0.025	216	31	101	0.31
25	152.4	0.009	79	12	72	0.17
30	155.5	0.002	21	3	34	0.10
35	0.0	0.000	4	0	11	0.00
40	0.0	0.000	1	0	2	0.00
Total			1,712	83	339	0.15125

Note: Power curve corrected to sea level. Accuracy +/– 5%.

Data from Gipe, P. (2009). *Wind Energy Basics*. White River Junction, VT: Chelsea Green Publishing Company.

Inflated Wind Speed Estimates One reason to explain the inaccuracy of using rated power as an energy output indicator is that wind speed estimates are often inflated. As mentioned earlier, the average wind speed for home sites is between 6 and 15 mph. Now consider that many industry experts list rated wind speeds 25 mph and higher, up to 70 mph.

Variations in Wind Speed Another reason rated power is inaccurate is that it gives one moment: the peak found on a wide-ranging scale. In other words, steady winds at the rated power figure or the maximum potential power are impossible.

Manufacturer's Estimates

Manufacturers usually base their estimates on a Rayleigh distribution, which is a common probability distribution. Estimates are also based at sea level. They are presented either in chart or table form. Manufacturers of small wind turbines usually provide average monthly generation estimates (see **TABLE 6-5**).

TABLE 6-5 ESTIMATES UNDER STANDARD CONDITIONS FOR SOUTHWEST WINDPOWER'S AIR BREEZE

Wind speed		AEO		
mph	m/s	~kWh/yr (measured)	% efficiency	kWh/yr (advertised)
10	4.5	277	28	264
12	5.4	418	25	444
14	6.2	537	20	600

Data from Gipe, P. (2009). *Wind Energy Basics*. White River Junction, VT: Chelsea Green Publishing Company.

In contrast, estimates from large wind turbine manufacturers provide annual averages using the AEO or AEP. Note, too, that these estimates are based on the assumption that wind direction is more or less stable. If wind direction changes frequently, these estimates do not apply especially for horizontal axis wind turbines (HAWTs). With frequent changes in wind direction, HAWTs become less efficient because wind flow would be off-axis and potentially obstructed by turbulence from other turbines, buildings, trees, and so on.

THE AMERICAN WIND ENERGY ASSOCIATION

If you've done online research regarding wind technology, chances are the American Wind Energy Association (http://www.awea.org/) has come up in most, if not all, of your searches. The association, commonly known as the AWEA, is the go-to source for most things related to wind technology. As its website says, this national organization is composed of more than 2,500 "wind power project developers, equipment suppliers, services providers, parts manufacturers, utilities, researchers, and others involved in the wind industry . . . [and] represents hundreds of wind energy advocates from around the world." When you access the AWEA's website, you can be guaranteed that the information is accurate and in touch with the newest developments in the field. Although the AWEA is a national organization, it also reports on and collaborates with the international wind industry.

The AWEA is there for both the novice and the professional. Its website offers basic information about wind technology for domestic use. For the larger scale projects in the industry, the AWEA gives valid information about developments in technology, current and upcoming wind energy projects, and companies in the field. What's more, the AWEA plays a key role in the drafting and passing of new laws related to renewable energy in the United States. It's safe to say that when you come across facts and figures relating to wind energy, the AWEA is probably at the source.

CHAPTER SUMMARY

Although there are three ways to estimate performance of a wind turbine—the swept area method, the power curve method, and use of manufacturers' estimates—always remember that these are estimates and not predictions. The swept area method is a quick way to get an idea of how much energy a turbine might generate. The power curve method provides a rough idea of the possible yearly energy generation. Manufacturer's estimates give you information about how a particular turbine might perform. It is best to view wind turbine estimates with a touch of skepticism. No matter how exact experts would like to be in their estimates, the wind is too variable a resource for such accuracy.

KEY CONCEPTS AND TERMS

Average wind speed
Instantaneous wind speed
Peak wind speed
Rated power

CHAPTER ASSESSMENT: POWER OUTPUT AND ENERGY YIELD

1. Energy describes the accumulation of power over a particular time period.
   - ❑ **A.** True
   - ❑ **B.** False
2. Because energy is usually measured in kilowatts and power is often described in watts, by how much do you divide power to determine kilowatts?
   - ❑ **A.** 10
   - ❑ **B.** 100
   - ❑ **C.** 1,000
   - ❑ **D.** 10,000
3. Small wind turbine manufacturers usually provide ________ estimates.
   - ❑ **A.** daily
   - ❑ **B.** weekly
   - ❑ **C.** monthly
   - ❑ **D.** annual
4. What kind of estimates do large wind turbine manufacturers provide? (Select two).
   - ❑ **A.** AEO
   - ❑ **B.** AES
   - ❑ **C.** AEP
   - ❑ **D.** AEA

5. What are the problems with using rated power? (Select three.)
   - ❑ A. It does not include instantaneous wind speed.
   - ❑ B. Wind is variable.
   - ❑ C. Rated power is not constant.
   - ❑ D. Rated power is constant.
   - ❑ E. Wind speed estimates are often inflated.
6. Instantaneous wind speed is the most important measure of wind.
   - ❑ A. True
   - ❑ B. False
7. What does the equation $P \times t$ calculate?
   - ❑ A. AEO
   - ❑ B. AEP
   - ❑ C. The amount of energy in the wind stream
   - ❑ D. The amount of energy produced
   - ❑ E. A and B only
8. Power measures the amount of energy in the moment.
   - ❑ A. True
   - ❑ B. False
9. How is energy measured?
   - ❑ A. Watt-hours
   - ❑ B. Watts
   - ❑ C. cm^3
   - ❑ D. m^2
10. In order to estimate how much energy is available in the wind over a particular time, time (t) is represented in:
   - ❑ A. minutes.
   - ❑ B. hours.
   - ❑ C. days.
   - ❑ D. months.

Commercial Applications

chapter 7

THANKS TO ADVANCES IN wind technology, electrical energy generated by wind turbines does not need to be used where it is generated. It hasn't always been this way. In the past, the norm was to use wind energy in the same location where it was generated. Today, wind turbines are used almost exclusively to generate electrical energy, which can be put to many different uses in various locations. In this chapter you discover three common commercial applications of wind energy: residential heating, water pumping, and desalination of salt water. These applications are followed by an examination of wind turbines that are interconnected with large utility grids. The chapter closes with a look at wind farms.

Chapter Topics

This chapter covers the following topics and concepts:

- Various standalone commercial applications of wind technology
- Wind turbines interconnected with large utility grids
- Wind farms and wind parks

Chapter Goals

When you complete this chapter, you will be able to:

- Describe the three primary standalone commercial applications of wind technology
- Outline the types of power storage and related issues
- List the advantages of having wind turbines interconnected with a large utility grid
- Describe wind farms/wind parks

Standalone Applications

In the early days of wind technology, **direct current (DC)** was commonly used. This allowed the generation and use of electrical energy at locations that weren't connected to the grid. When the amount of energy is limited to a low number of watts, such as 300 or 400, direct current is sufficient. Today, however, direct current has been replaced by **alternating current (AC)**, which can handle the generation of a high number of watts but does not provide a constant current necessary to maintain output. As a result, today's turbines that function independently need supportive technology to ensure success.

Wind turbines capable of blade pitch control can help provide a somewhat constant current, but this is not enough. The most important element in successful standalone applications is the existence of an energy storage system. Usually batteries are used as the storage device but this option generally proves too costly. In the odd case, water can be an efficient means of storing energy. Read on to discover more about power supply in standalone applications and the issues associated with energy storage.

Power Supply

As you know, wind is not constant. Therefore, when the wind does not provide the necessary power to meet the output expectations of a turbine, a backup mechanism is imperative. This backup comes in the form of energy storage. There are various methods of energy storage.

The methods of energy storage most commonly referred to include:

- Direct storage of mechanical energy
- The storage of converted electrical energy in batteries
- The use of hydrogen
- The use of water

Next, you will find a brief description of each method and its challenges.

Storage Issues

The most obvious issues with storage are related to capacity and cost. Today's methods cannot provide huge amounts of storage. Besides, all the options are expensive.

The advantage of storing mechanical energy directly is that the energy need not be converted because it is compatible with the final user's needs. Often, when you have to convert energy into another form to store it, you lose energy. One option that has been explored is storing mechanical energy in flywheels. Flywheels are spinning mechanical objects that are able to capture and accumulate energy until needed. Wind technology investigators have explored various flywheel designs. They have found that high-speed flywheels can absorb and store a great deal of energy, but the mechanics behind this technology are too complicated. Researchers found that the more traditional, slower flywheels cannot store nearly as much energy. In addition, the slower flywheels also experience significant energy losses. Flywheels are best used for short-term storage and when operating with small grids.

Another method of energy storage is through the use of batteries. For the time being, utilizing batteries as a storage device for electrical energy is limited to short-term use. Neither the traditional battery made of nickel and cadmium nor the more advanced designs made of silver and zinc have the capacity for storing a lot of energy over a long period of time.

Another option for storing electrical energy that is presently attracting attention is through the generation of hydrogen. There are various reasons this option is being explored. For one, it is environmentally friendly. Additionally, it can be stored at low temperatures. It can be stored as well in its original form as a gas or it can be chemically absorbed into another appropriate form. Despite these advantages, more research is required before this storage method can be applied on a large scale.

The last option described is using the original method of energy storage: water. It is one storage method that can be used for small- and large-scale applications. Examples include water pumps, desalination of water, and small-scale heating. Water as a storage method can be an affordable option when the topographical features around the water source are suitable.

Residential Heating

The first commercial application of wind energy examined in this chapter is residential heating. Consider that more than three-quarters of energy costs in countries such as Germany originate in space heating. The financial and environmental gains could be great if wind energy covered these energy needs. It is for these reasons that many people advocate using wind energy for this purpose. What research has shown, however, is that relying on just wind power for heating purposes is insufficient. For wind power to be utilized in an affordable manner it needs to be paired with a conventional heating system. The other option is to add a small energy-storage device to an existing heating system. In this way, wind energy can be stored and used as needed.

Note that when oil or gas prices are extremely low, residential heating by means of wind energy is not economically feasible. For example, in 2003 the cost of heating energy from a gas or oil source was about 3 cents/kWh. To pair a traditional heating system with a wind energy storage system would cost more than the potential savings. But with higher gas or oil prices, using wind energy, especially with tiny wind turbines, may bring positive returns.

Water Pumping

Using wind power to pump water is one of the oldest applications of wind power in America and Australia. It used to be that small quantities of water were accessed from deep locations for drinking water. In contrast, today large quantities of water are needed to support agriculture, especially in poorer countries. Additionally, this water frequently is pumped from shallow ground. Due to environmental conditions, it is common to see the wind turbine and the water pump in two separate locations. Under these circumstances, a combination of a diesel generator and wind turbine is often used.

Wind turbines can operate in conjunction with two types of water pumps:

- **Piston pump**—has an approximate 80 to 90 percent efficiency rate and can maintain good efficiency at slower rotational speeds
- **Centrifugal pump**—has an approximate 50 to 75 percent efficiency rate. Efficiency drops with decreasing speed.

Despite the centrifugal pump's lower efficiency rate, when it is matched with a high-speed wind turbine, it actually performs better than the piston pump **FIGURE 7-1**. This is because the centrifugal pump and the wind turbine share many of the same operating traits.

At present, using wind power for pumping water is too expensive for developing countries. But as its popularity grows and prices drop, experts predict that wind power for pumping water will be successfully used throughout the world.

Desalination of Sea Water

If you think that knowing how to separate salt from water, or desalination, isn't necessary, think again. Predictions are that drinkable water will be a scarce commodity in the future. Making use of the vast amounts of ocean water seems to be the remedy. In fact, sailors knew how to separate salt from sea water on ships in the early 1800s.

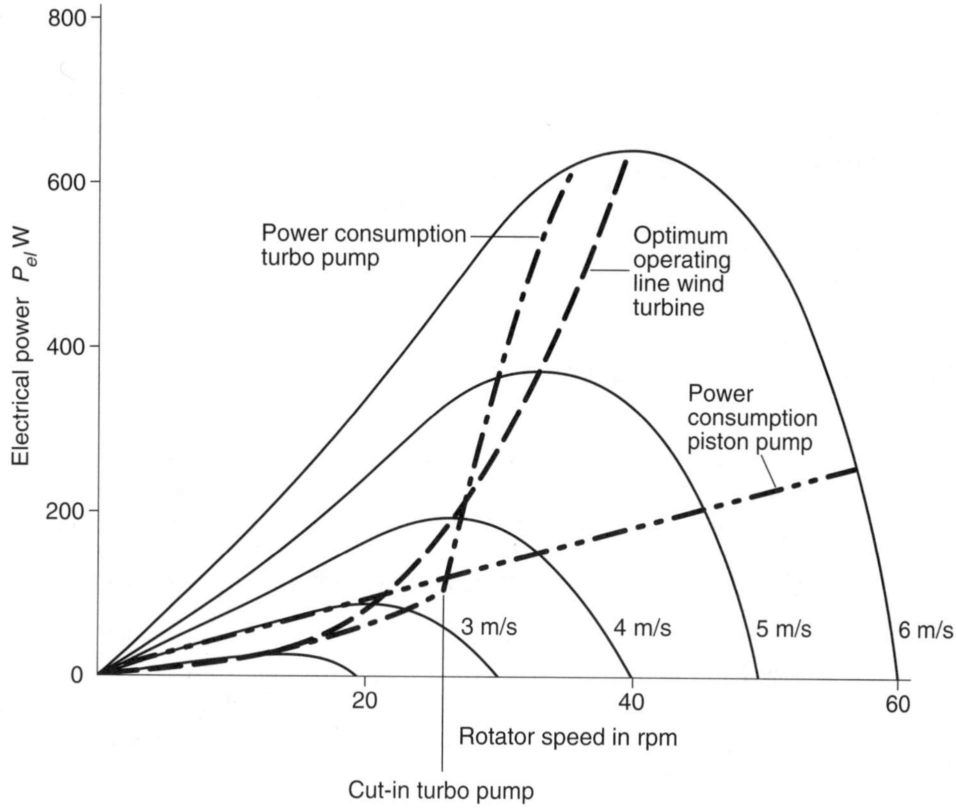

FIGURE 7-1 A comparison of the centrifugal (turbo) and piston pumps with the ideal conditions of a wind turbine.
Adapted from Hau, E. (2006). *Wind Turbines*. Berlin and Heidelberg: Springer-Verlag.

Desalination can be achieved in two ways:

- **Distillation**—Using thermal energy, this method can remove almost all salt from seawater. A plant that uses 150 megawatts can produce 30,000 to 40,000 m^3 of drinking water.
- **Membrane methods**—Reverse osmosis (RO) **FIGURE 7-2** is the most common of the membrane methods. This process involves a salt-water container and a fresh-water container, divided by a membrane. Salt water is channeled into its container at a higher pressure than the fresh water. This results in some of the water filtering through the membrane, into the fresh water, and then into storage. Whatever salt remains is then removed. Currently, brackish water, or water with a low salt content, is the easiest to desalinate. As the salt content increases, so does the energy required for the pump.

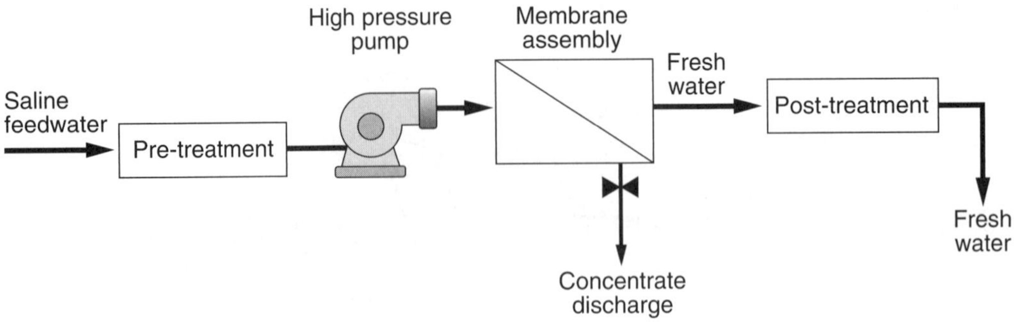

FIGURE 7-2 A simple diagram illustrating reverse osmosis.

Of the two methods, RO is the preferred one when wind energy is used. **FIGURE 7-3** shows you the layout of a test plant. With an average wind speed of 7 m/s and salt water with 36 g/kg of salt, about 13 m^3 of drinking water can be produced.

Wind Turbines Interconnected With Large Utility Grid

If you consider how much wind energy potential exists throughout the world, you find that almost all of this capacity is connected to large utility grids. Why? Here are three good reasons: One, power output functions independently of the immediate energy needs of the consumer. Two, traditional power stations provide support when output does not meet demand. And three, rotor speed can be managed by the power station.

The first country to connect wind turbines to the public grid was Denmark. In 1978, Danish farmers and small business owners interconnected their wind turbines to the larger utility grid **FIGURE 7-4**. They had the support of the federal government and received lower electricity costs. Around 2007, Germany, Britain, the Netherlands, and Spain each had approximately 15,000 wind turbines operating under the control of the public grid. The turbines are connected to the fixed-frequency of the grid, which ultimately manages the speed. You can also say that the wind turbines are operating in parallel with the grid. Currently, a meter is like a bridge between the turbine and the grid.

Wind Farms or Wind Parks

Wind farms, also referred to as wind parks, are areas that contain several wind turbines organized spatially. Wind farms make economic sense. When extensive distances exist between the turbine and an in-feed location, it is only logical to connect that same feed to many turbines and not just one. Each turbine may or may not be connected to other turbines but the turbines are all connected or in parallel to the primary grid.

Wind farms began in the early 1980s in the state of California. Danish manufacturers had control of the market until 1986, when US federal tax benefits for

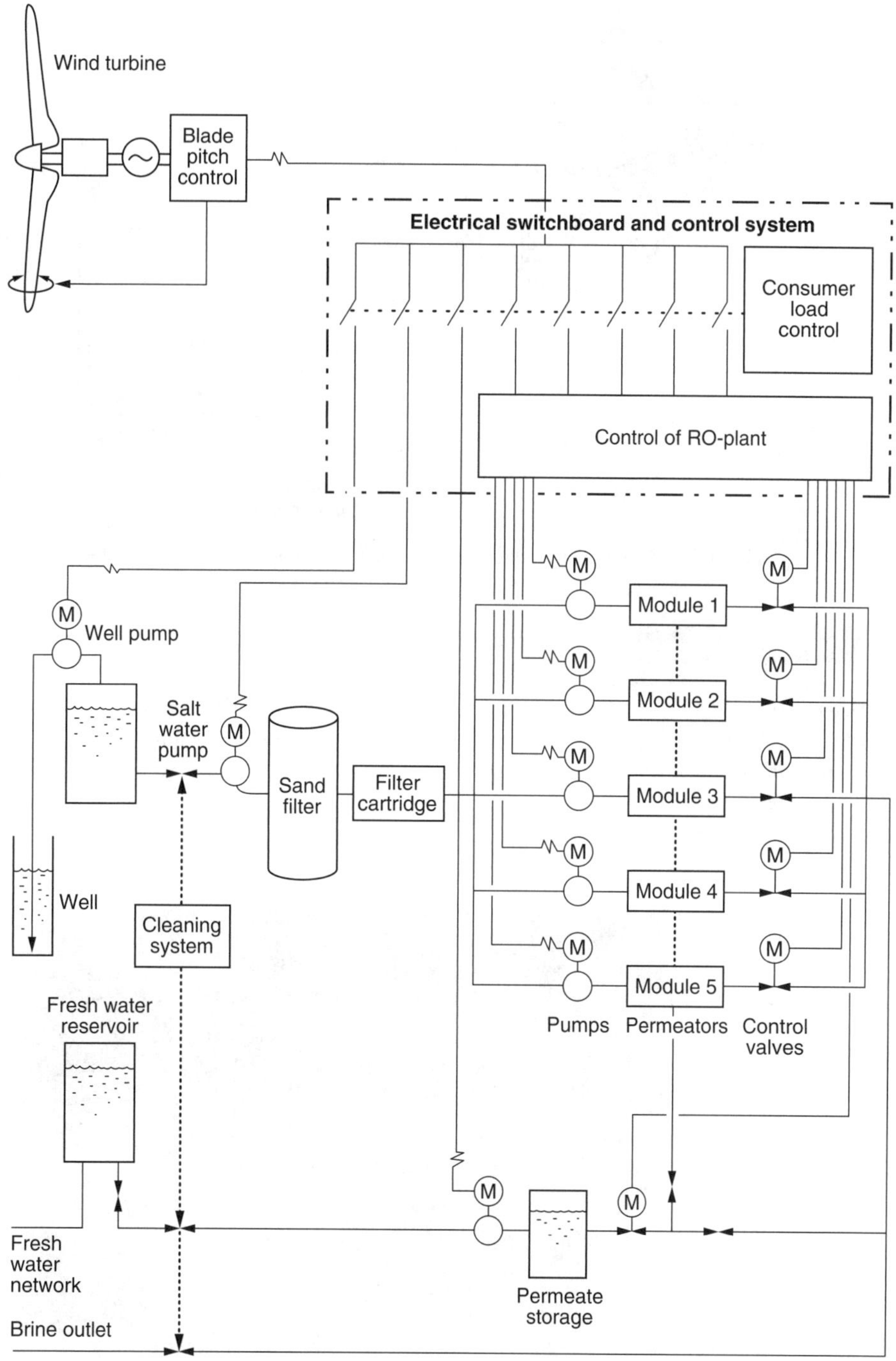

FIGURE 7-3 The layout of a test plant that combines wind energy and reverse osmosis.

Adapted from Hau, E. (2006). *Wind Turbines*. Berlin and Heidelberg: Springer-Verlag.

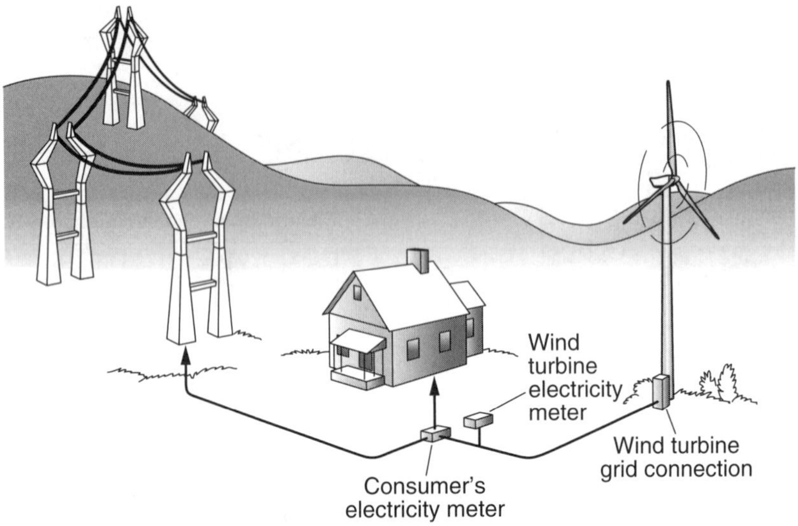

FIGURE 7-4 A residence that uses wind energy and is interconnected with the larger utility grid.
Adapted from Hau, E. (2006). *Wind Turbines*. Berlin and Heidelberg: Springer-Verlag.

NOTE

Although the terms "wind farm" and "wind park" are synonymous, wind farm is the more widely used term. In Europe, "wind park" may refer a small group of turbines organized in a specific configuration.

alternative energy expired. By 1990, Germany led the way for bigger wind farms. In the earlier days, wind farms were located only in coastal regions. But today, many wind parks can be found in the interior regions of Europe and the United States **FIGURE 7-5**. In Europe, Spain stands out with wind parks that produce several hundred megawatts of energy. Another type of wind farm is an offshore farm, which is growing in popularity.

FIGURE 7-5 In the past, wind parks were located only in coastal regions. Today they can also be found in the interior regions of Europe and the United States.

SMALL WIND CERTIFICATION

Although wind turbines have been in use for some time now, a certification process has come into effect only as recently as 2010. The Small Wind Certification Council (SWCC) regulates the small wind turbine industry. The American Wind Energy Association (AWEA) has a set of standards known as Small Wind Turbine Performance and Safety Standards, which the SWCC enforces. To get an idea of how rigorous the certification process can be, consider that testing takes a minimum of six months and up to two years. Note that the SWCC does not actually perform the tests. Instead, the organization checks and certifies the results of tests carried out by third parties.

Certification plays more than one role. The most obvious is that it ensures certified wind turbines meet approved minimums. Another function is that it provides governments and utility companies, which can provide rebates and incentives, a standard against which to measure turbines. One of the reasons the SWCC was established is that power ratings of small turbines differed by 40 percent. Other reasons included a lack of understanding by consumers and unsafe or inadequate products in use.

The wind energy community has applauded the founding of the SWCC. The very first wind turbines began the testing and certification process as of June 16, 2010. The wind energy industry, both technology and usage, is bound to improve through this process.

CHAPTER SUMMARY

An examination of the commercial applications of wind energy revealed that storage is the most important issue. All of the four primary methods of energy storage face high costs and limited capacity. A look at the three main commercial applications of wind energy showed how integral wind energy is to our future. Wind turbines and wind farms connected to the large utility grid have the potential to meet future global needs. As natural resources continue to be depleted, wind energy shows tremendous promise for both large and small projects.

KEY CONCEPTS AND TERMS

Alternating current (AC)
Direct current (DC)

CHAPTER ASSESSMENT: COMMERCIAL APPLICATIONS

1. For wind power to be utilized in an affordable manner it needs to be paired with a conventional heating system.
   - ❑ A. True
   - ❑ B. False
2. The process of ________ removes the most salt from seawater.
   - ❑ A. semi-desalination
   - ❑ B. distillation
   - ❑ C. reverse osmosis
   - ❑ D. membrane separation
3. How much wind energy capacity is connected to large utility grids?
   - ❑ A. Half
   - ❑ B. Three-quarters
   - ❑ C. Almost all of it
   - ❑ D. All of it
4. Batteries as an energy storage device are best limited to ________ use.
   - ❑ A. short-term
   - ❑ B. moderate
   - ❑ C. long-term
   - ❑ D. occasional and limited
5. In developing countries, water is usually pumped from great depths.
   - ❑ A. True
   - ❑ B. False

6. What are the two primary concerns with wind energy storage? (Select two.)
   - ❑ A. Capacity
   - ❑ B. Sustainability
   - ❑ C. Environmental impact
   - ❑ D. Cost

7. In which ways can wind energy be stored? (Select three.)
   - ❑ A. Water
   - ❑ B. Carbon dioxide
   - ❑ C. Hydrogen
   - ❑ D. Batteries
   - ❑ E. Electromagnetism

8. To say that the wind turbines are operating in parallel with the grid means that the larger grid controls speed.
   - ❑ A. True
   - ❑ B. False

9. Which storage method can be used for small and large applications?
   - ❑ A. Hydrogen
   - ❑ B. Water
   - ❑ C. Batteries
   - ❑ D. A and B only
   - ❑ E. None of the above

10. In Germany, more than three-quarters of energy costs come from:
   - ❑ A. lighting.
   - ❑ B. water heating.
   - ❑ C. space heating.
   - ❑ D. None of the above

Offshore Wind Energy

chapter 8

THE AMOUNT OF LAND on the planet is finite. Specific environmental and geographical conditions are also required to support a wind farm on land. This further narrows the selection of suitable space for new wind farms. By contrast, the scope of available space offshore is immense. When you combine this reality with the fact that wind speeds are higher over the sea, it makes sense that offshore wind energy is growing in popularity. An examination of the factors related to offshore wind energy provides you with a general understanding of the topic. The chapter also explores the characteristics of prospective sites and the types of structures that are in use today.

Chapter Topics

This chapter covers the following topics and concepts:

- What factors are related to offshore wind energy
- What the basic characteristics are for the support structures for offshore wind turbines
- What types of structures are used for offshore wind turbines

Chapter Goals

When you complete this chapter, you will be able to:

- Outline the various considerations related to offshore wind energy
- Describe the basic characteristics found in support structures for offshore wind turbines
- Describe the types of structures used for offshore wind turbines

Offshore Wind Energy Considerations

Wind energy on land and wind energy offshore are not the same. In fact, there are many unique factors to consider when looking at offshore wind energy. The following sections explore these factors.

Wind Resource

Although wind is not constant, it can be said that wind speed over the sea is consistently higher and more stable than wind over land. The main reason for this is the amount of surface roughness. Naturally, there is a higher surface roughness on land due to landscape, manmade structures, etc. Over sea, the lower surface roughness facilitates higher wind speeds.

Characteristics of Prospective Sites

Knowing the ideal traits of potential sites for offshore wind farms is crucial in the planning stages. European offshore wind farms can generate up to 40 percent more energy than on land, but only if the turbines sit at about 10 kilometers from shore **FIGURE 8-1**. A few kilometers less or a few more could produce a decreased energy yield. Of course, other environmental factors need to be considered. For example, the coastline of the Baltic Sea does not contain the typical tidal shallows and mud flats other coastlines in the region possess. In this particular case, offshore sites can function well alongside the coast.

Other factors must be considered when choosing a prospective site. The following three sections examine water depth, the type of soil found on the sea bottom, and the issues surrounding the laying of electrical cables.

Water Depth

Experts agree that water depth is the most important factor for choosing the right location for an offshore turbine. There is not one specific water depth that is ideal for all offshore turbines. Instead, a range of depths is offered for the various types of structures, but as a rule, larger turbines require deeper water. Water depth can range

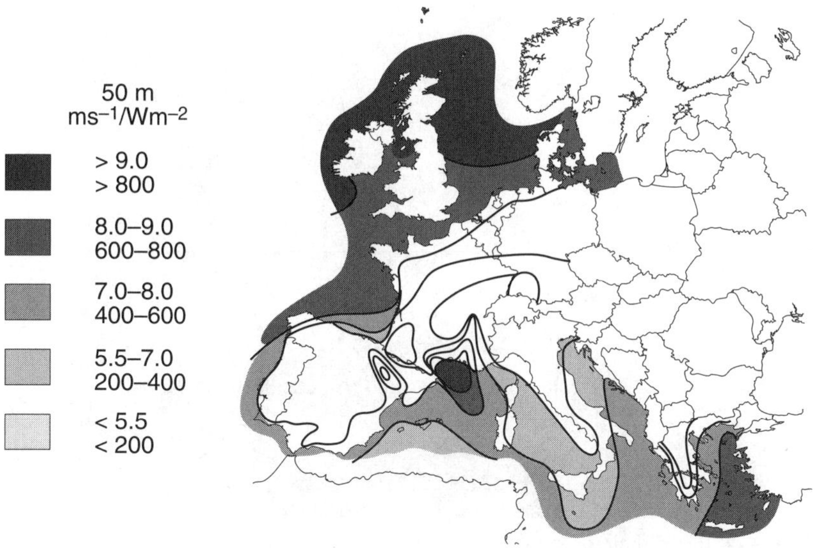

FIGURE 8-1 This figure illustrates the wind conditions of the European coastline starting at 10 km from the coast and at five different heights.
Data from Hau, E. (2006). *Wind Turbines.* Berlin and Heidelberg: Springer-Verlag.

from 5 to 30 meters. Experimental offshore turbines can be built in depths from 50 to a maximum of 100 meters. The latest designs are floating platforms that can be positioned in much greater water depths.

Nature of Soil at Sea Floor

The type of soil that makes up the sea floor plays a significant role when choosing a potential site. Extensive soil testing is mandatory to determine if the sea bottom is suitable and strong enough to support an enormous wind turbine. Research also involves the investigation of the currents, because they have the power to displace soil and sand, which changes the ground's conditions. Furthermore, ocean currents often have a "scouring" effect on foundations, so they must be avoided.

Placing Electrical Cables

When you compare cable systems and grid connections for turbines on land and offshore turbines, you see that offshore wind farms are much more complicated. In comparison to electrical systems for land turbines, manufacturers of offshore turbines have to be particularly aware of creating reliable (redundant) electrical systems, the increased costs of offshore electrical systems, and the distance between the turbine and land.

Four areas need to be considered:

- **Internal electrical system**—This consists of a three-phase system that ranges from 20 to 40 kV. The plastic-covered sea cables are much more costly

FIGURE 8-2 The offshore substation of the Rødsand/Nysted wind park in Denmark.

than land cables (about 20 to 40 percent more expensive). What increases costs further is the laying of the sea cables, which accounts for approximately 50 to 80 percent of the costs. Specialized ships are used to lay the cables; they push the cables one meter below the sea floor using a water jet.

- **Offshore transformer station**—Offshore turbines that are located a fair distance from land and use higher voltages have their own transformer at the turbine site that connects to a central transformer substation **FIGURE 8-2**. The extra transformer necessitates an offshore substation. Note that higher voltages range between 110 and 150 kV.
- **Cable connection to land**—When the wind farm is far offshore, it is harder for the alternating current to be transmitted through the cables. The maximum distance in which the cables can transmit the current is about 100 kilometers. Know that the efficiency of the cables decreases with length. Laying parallel cable lines is common **FIGURE 8-3**.
- **Connecting with grid on land**—As offshore wind farms increase in power, it becomes necessary to connect them to an extra-high voltage grid (using extra-high voltage transmission lines). Consider that the first offshore wind farms had powers of 100 or 200 MW. Future large offshore wind farms are predicted to have powers of 1,000 MW.

Wind Farm Design and Layout

Generally speaking, offshore wind turbines need to be positioned farther away from one another as compared to turbine layout on land **FIGURE 8-4**. This is because of a greater wake behind the rotor of an offshore turbine, which in turn

FIGURE 8-3 Workers laying cable at the Kentish Flats, an offshore wind farm in the United Kingdom.

© Chris Laurens/Alamy Images

FIGURE 8-4 An example of an offshore wind farm, where the wind turbines need to be positioned farther away from each other compared to their placement on land.

© Pics-xl/ShutterStock, Inc.

can negatively affect the functioning of turbines positioned downwind. Without this distance between them, overall energy yield will be reduced. Ideally, offshore wind turbines need to be placed approximately a half-mile apart (approximately 0.8 km) from each other.

Operation and Maintenance

One of the challenges of operation and maintenance is reaching the wind turbine site. When the water is rough, fixing of problems can be delayed, which can lead to economic loss. For instance, in the early days of offshore turbines, the Bockstigeen wind farm off the coast of Sweden experienced operational delays because the maintenance boat had trouble docking in the meter-high waves. Submarines are a future possibility for performing maintenance on offshore wind farms. Helicopters are another potential option but are also restricted in bad weather.

Due to the challenges that weather presents, preventive maintenance is important. Part of this kind of maintenance involves identifying potential failures. When the weather is calm, maintenance work is done with the ultimate goal of having a high mean time between failures (MTBF).

Environmental Issues

Offshore wind farms have less environmental impact than wind turbines on land. One common complaint against onshore turbines, usually from the tourist industry and environmentalists, is the disruption of the visual landscape. Regulations exist according to each country as to how far out an offshore turbine must sit. For example, Danish wind farms must be established between 6 and 12 kilometers from the coastline.

Noise is another environmental issue that presents problems for onshore turbines but not usually for offshore turbines. However, opponents say that the noise produced by turbines travels underwater and disturbs sea life. In contrast, advocates for offshore farms point out that motorized vehicles have been used on water for more than a century without causing major adverse effects. Research shows that the underwater noise is at a low frequency that many marine animals do not hear.

Construction of offshore farms disturbs sea life but, according to Danish research, it does not cause significant upset to the seabed or wave patterns. It should be noted, however, that current studies span six-month periods. The long-term effects on marine life have yet to be determined. According to the American Wind Energy Association, certain populations of fish are being rejuvenated because of offshore farms as the foundation provides a safe place for spawning. Some designers are looking at making their wind turbine foundations act as artificial reefs to help the aquatic ecosystem.

Another area of concern is bird life. Extensive testing shows that birds actually avoid the turbine sites. Also worth noting is that few birds fly below a height of 200 meters. It is advisable to investigate the breeding and migration patterns of local birds when choosing a site so as to mitigate any disturbance to them.

Wind Energy Economics

A common argument that supports the expansion of offshore wind farms is that the higher wind speeds available for use will translate to cost savings. Reality shows something different. Offshore wind farms commonly have higher building and operating costs, which can substantially reduce projected savings. As technology improves, larger and larger offshore turbines are being erected, which translates to increased energy generation and, ultimately, savings.

Turbulence

The difference between turbulence over land and over sea is notable. Turbulence intensity over land runs from 10 to 20 percent. Over the sea, it doesn't reach even 10 percent. This means that the turbine is less stressed, with a lower fatigue load. However, as mentioned earlier, the reduced turbulence also affects the wake behind the rotor, so that offshore turbines need to be placed farther apart than turbines on land.

Waves

The height of waves is of almost equal importance to the depth of the water. When there are high winds over the open sea, there will also be high waves. If a turbine is built in a site that experiences consistently high waves, maintenance costs will be higher. Average and peak wave heights need to be determined when assessing a site. The blades need to be able to clear the maximum wave height by a healthy margin.

Support Structures

The offshore wind turbine is composed of three basic parts: tower, substructure, and foundation.

Tower

Because wind speeds are higher over sea than land, offshore towers need not be nearly as high. In fact, the ideal tower height for maximum generation of energy is lower at sea. When determining the tower height, other facts such as water depth and wave height are considered.

Substructure

The substructure is what sits between the tower and the foundation. The substructure contains such areas as the work platform and the boat landing.

Foundation

A foundation that can support the entire turbine is of the utmost importance. It is not enough to duplicate the basic foundation used for a turbine on land.

Foundation types differ according to the structural design, reviewed in the following section. Take note that the costs for laying a foundation increase greatly as water depth and wave height increase.

Types of Structures for Offshore Wind Turbines

As mentioned earlier, harnessing wind energy on land and offshore requires different considerations. More than one type of structure exists for offshore wind turbines. The following outlines the most commonly used structures.

Monopile

The simplest of structures is the monopile, which is one long pipe inserted upright into the foundation **FIGURE 8-5**. If the sea bottom is made up of primarily sand or gravel, supporting the monopile is the most cost-effective option because drilling harder sea bottoms costs much more money. The pipe is most often made of steel and ranges between 4 and 6 meters in diameter. Current experts agree that monopile structures should be erected only in waters up to a maximum of 25 meters. This structure is not stable on stony seabeds.

Gravity Structures

Gravity structures **FIGURE 8-6** involve a **caisson** made out of concrete and filled with sand or gravel. Sometimes the caisson can be composed of steel to reduce the structure's overall weight. Before the caisson is filled, it is floated out to the offshore site. At this point it is weighted to the sea bottom. This type of structure is best suited for shallow water up to a maximum of 10 meters, which is limiting. A plus for environmentalists is that this structure can be removed with little disruption to the surrounding area.

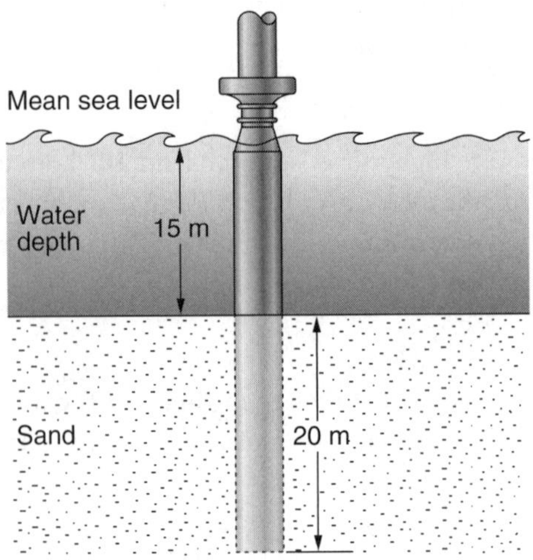

FIGURE 8-5 A diagram of a monopile structure.

Adapted from Hau, E. (2006). *Wind Turbines.* Berlin and Heidelberg: Springer-Verlag.

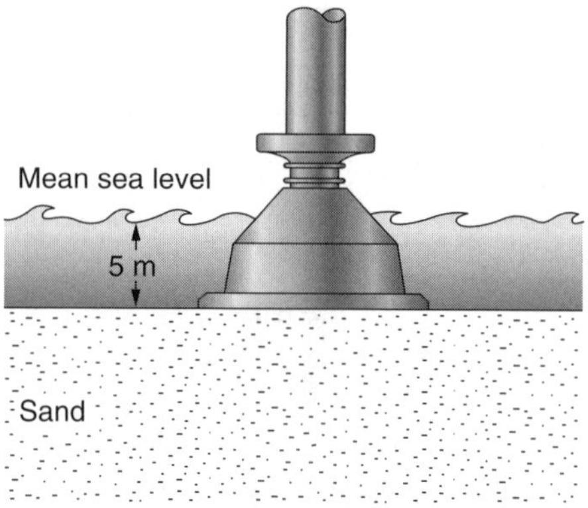

FIGURE 8-6 An illustration of a gravity-type structure.
Adapted from Hau, E. (2006). *Wind Turbines.* Berlin and Heidelberg: Springer-Verlag.

Multimember Support Structures

Offshore wind turbines situated in medium to deep waters require multimember support structures, such as the tripod and the jacket types.

Tripod

A tripod is a three-legged structure that contains a steel tube in its center **FIGURE 8-7**. Because it is light, it works well and is known to have high stability in

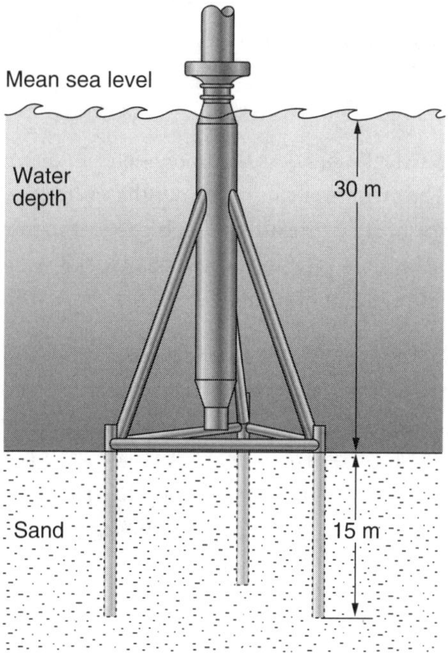

FIGURE 8-7 A sketch of a tripod support structure.
Adapted from Hau, E. (2006). *Wind Turbines.* Berlin and Heidelberg: Springer-Verlag.

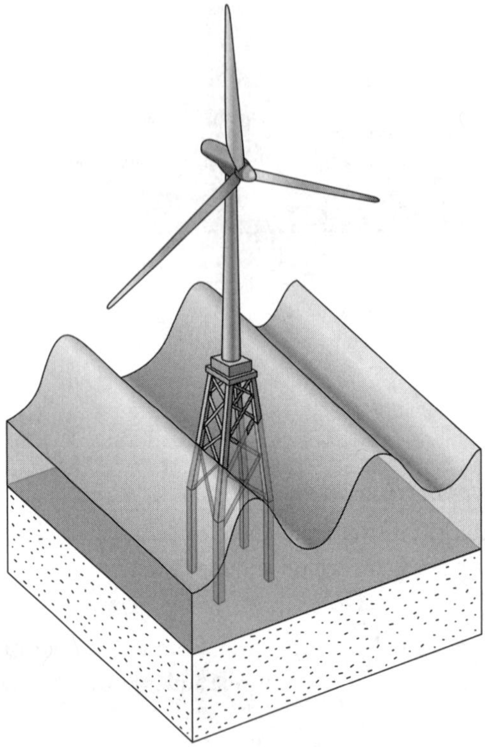

FIGURE 8-8 A diagram of a four-legged jacket structure.
Adapted from Deepwater Wind

deeper waters. Smaller steel tubes, less than one meter in diameter, secure each leg to the sea bottom. The structure can reach depths up to 20 meters. The biggest drawbacks to this type of structure are the high costs of manufacturing it on land and the issue of transportation. The tripod can be made of steel or concrete. Similar to the monopile, the tripod cannot be used when the sea bottom is stony.

Jacket

The jacket structure is a design that is already popular in the oil drilling industry. It has four legs that are anchored to the sea bottom by piles between 0.8 and 2.5 meters in diameter **FIGURE 8-8**. An advantage to the jacket structure is that is uses 40 to 50 percent less steel than the monopile structure.

THE PROS AND CONS OF OFFSHORE WIND FARMS

Which side of the fence are you on? Are you for offshore wind farms or against them? Or are you still undecided? Take a look at the pros and the cons of offshore wind farms to make an informed decision.

The pros:

- Offshore wind has minimal environmental impact.
- Bigger wind farms can be constructed offshore than on land.
- Wind is free and easily accessed, unlike resources controlled by other countries.
- Great potential exists to reduce negative environmental effects.
- Offshore turbines generally are not noticeable from land, preserving the natural landscape.
- Bothersome noise that disturbs residents on land are not heard offshore, and its low frequency doesn't affect marine life.

The cons:

- Construction, transportation, operation, and maintenance costs are often high.
- The wind is variable.
- The long-term environmental effects on sea life and weather patterns are presently undetermined.

Chapter Summary

As of 2009, a total of eight countries had offshore wind farms in use: Denmark, Belgium, Sweden, Finland, Britain, Ireland, Germany, and the Netherlands. By 2015, five additional countries from the European Union will have offshore farms. Although offshore farms have not been in use for long, they are clearly the wave of the future. This chapter provided you with an overview of offshore farms, including the ideal traits of potential turbine sites, how offshore farms should be configured, issues related to operation and maintenance, environmental concerns, and financial issues. As well, this chapter gave you the basic information related to the main types of offshore structures currently in use today.

Key Concepts and Terms

Caisson

Chapter Assessment: Offshore Wind Energy

1. Turbulence intensity over the sea is less than 10 percent.
   - ❑ A. True
   - ❑ B. False
2. What is the ideal approximate distance between offshore wind turbines in a wind farm?
   - ❑ A. 50 meters
   - ❑ B. 100 meters
   - ❑ C. Half a mile
   - ❑ D. Half a kilometer
3. Gravity structures are best suited to maximum water depths of ______ meters.
   - ❑ A. 10
   - ❑ B. 15
   - ❑ C. 20
   - ❑ D. 25
4. Which type of support structures are multimember structures? (Select two.)
   - ❑ A. Monopile
   - ❑ B. Tripod
   - ❑ C. Gravity
   - ❑ D. Jacket

5. Generally speaking, the costs for laying a foundation decrease as water depth and wave height increase.
   - ❑ A. True
   - ❑ B. False

6. The greatest obstacle(s) to offshore turbine maintenance is/are:
   - ❑ A. water depth.
   - ❑ B. weather conditions.
   - ❑ C. wave height.
   - ❑ D. Both A and B

7. Offshore wind farms have more of an environmental impact than wind turbines on land.
   - ❑ A. True
   - ❑ B. False

8. When the seabed is stony, which support structures cannot be used? (Select two).
   - ❑ A. Monopile
   - ❑ B. Tripod
   - ❑ C. Gravity
   - ❑ D. Jacket

9. The maximum distance of cables that connect the current from land to the offshore site is approximately ________ kilometers.
   - ❑ A. 10
   - ❑ B. 70
   - ❑ C. 100
   - ❑ D. 140

10. Lower surface roughness over the sea results in higher and more stable winds than over land.
    - ❑ A. True
    - ❑ B. False

11. Which of the following are the two most important factors when assessing a site for an offshore turbine? (Select two.)
    - ❑ A. Water depth
    - ❑ B. Existing marine life
    - ❑ C. Wave height
    - ❑ D. Turbulence

12. The most cost-effective support structure is the ________ structure.
    - ❑ A. jacket
    - ❑ B. gravity
    - ❑ C. tripod
    - ❑ D. monopile

chapter 9

Wind Turbine Costs and Economic Analysis

A QUICK LOOK AT the costs associated with wind turbines shows they are not cheap. Analyzing the costs of wind turbines involves an examination of various factors. This chapter explores the costs of wind turbines as well as the issues related to those costs. It includes an exploration of the economics related to wind turbines connected to the grid as well as a look at economic approaches to wind technology.

Chapter Topics

This chapter covers the following topics and concepts:

- What the economics and costs are for wind turbines connected to the grid
- How the economics of wind turbines are analyzed

Chapter Goals

When you complete this chapter, you will be able to:

- Describe the factors and costs related to wind turbines connected to the grid
- Explain how the economics of wind turbines are analyzed

Wind Turbines Connected to the Grid

For the last few decades, the popularity of wind turbines has been linked to the price of oil. Interest in wind turbine technology soared during the oil crisis in the 1970s. People feared that the high prices of oil would rise even higher and make electricity too expensive to consume. This uncertainty resulted in consumers connecting small wind turbines to the grid via a kilowatt-hour meter in order to reduce their electricity bills **FIGURE 9-1**. Obviously this meant less money for electric utilities. Naturally, these companies did what they could to impede the interconnection.

In 1978, the US Congress passed the Public Utility Regulatory Policies Act (PURPA), which made interconnection easier. Additionally, subsidies were provided as incentives to make use of wind technology. More than 4,000 wind turbines popped up throughout North America, but their popularity was short-lived. The main culprit was the drastic crash of oil prices. Electricity became cheap again, so wind turbines were returned to their original purposes: pumping water and charging batteries. In Europe, however, Germany and Denmark continued to employ interconnection with the grid. North America learned about connecting to the grid by their example.

Today, as long as electricity prices remain high, wind turbine technology maintains its appeal. And not just for small wind turbines either. At present, large and medium-sized wind turbines are used commonly in wind farms and continue to grow in popularity. Large wind farm plants generate electricity for sale to other cities. Individual large wind turbines and small groups of large turbines also make use of interconnection to sell electricity locally.

Energy Capture Efficiency

To determine the amount of energy captured and the effectiveness of wind turbines connected to the grid, look at the technology. Commonly, a wind turbine

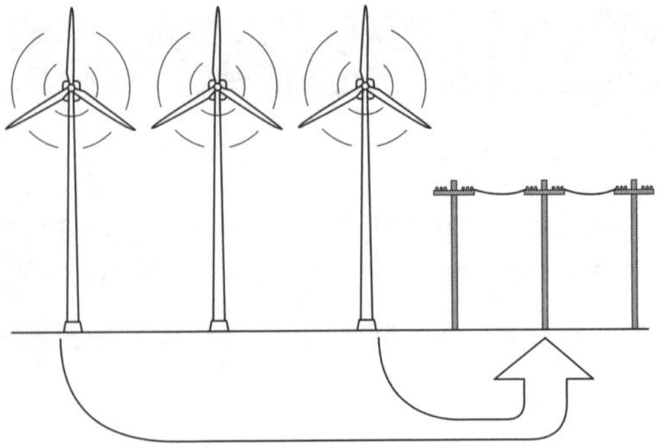

FIGURE 9-1 This figure illustrates the concept of selling electricity produced by wind turbines for ultimate sale to the grid.

Adapted from Gipe, P. (2009). *Wind Energy Basics*. White River Junction, VT: Chelsea Green Publishing Company.

uses an **induction generator**, also known as an **asynchronous generator**. Such generators are, by far, the easiest to connect to the grid. The current from the grid is used to magnetize the field within the generator. The drawback to this system is that when the grid is down, the generator cannot function. This situation obviously reduces the efficiency of the entire system. Induction wind turbines employ internal controls that manage the connection to the grid. When the grid is down, these controls tell the turbine to disconnect. Without such a mechanism, the turbine connected to the grid would actually use the available electricity. An inverter is often combined with a generator to generate electricity that is compatible with the grid. In fact, the most common option today is the pairing of generators with inverters.

System Availability

As mentioned, utility systems weren't thrilled with the passage of PURPA in the United States, and they are still impeding connections despite formal regulations against this. Sometimes, such cases end up as lawsuits, even though worldwide, wind turbines have functioned with the grid for over 10 billion hours. In other developed countries, such as Canada or Mexico, no such legislation exists. In Europe, however, there is much more cooperation. Germany in particular stands out. This country passed the Renewable Energy Sources Act. It prescribes that, if safety and quality standards are met, the utilities must choose renewable energy sources over all others.

Much argument surrounds **net metering**, an important concept in the United States. As mentioned, a meter sits between the turbine and the grid **FIGURE 9-2**. When the consumer uses energy from the grid, the meter spins forward. When the consumer sends energy back to the grid, the meter spins backward. The concept is called net metering because the customer pays the net difference between the two. Net metering is beneficial for customers because in some US states,

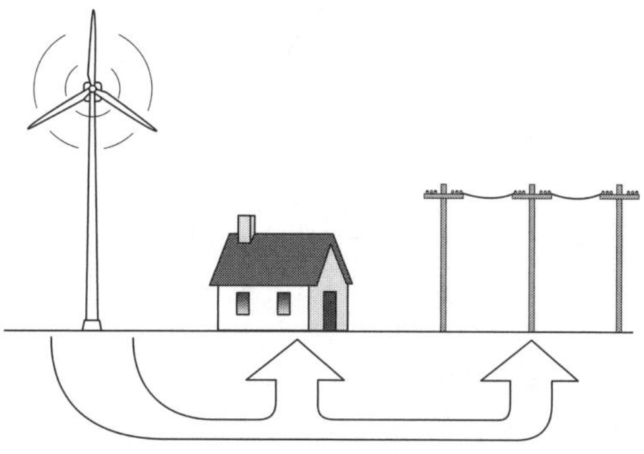

FIGURE 9-2 The basics of net metering.

Adapted from Gipe, P. (2009). *Wind Energy Basics*. White River Junction, VT: Chelsea Green Publishing Company.

customers can carry over energy they don't use, or even sell the energy back. The majority of states have laws related to net metering, but in some places net metering remains illegal.

System Lifespan

It is important to know the lifespan of a wind turbine to decide whether the investment is worth it. Not all turbines have the same lifespan. A turbine can last for a few years or for decades. It is necessary to examine the various sizes of turbines and their longevity separately:

- **Micro and mini wind turbines**—Experience has shown wind technology experts that these last from 6 to a maximum of 12 years. Generally speaking, the more lightweight the turbine, the shorter its life. Because some turbines are expected to last only a short time, owners are less likely to repair them.
- **Household-sized wind turbines**—Estimates calculate that they have a life of 15 to 20 years. Note that in order to have such a long life, the turbine must receive regular maintenance and care.
- **Small commercial wind turbines**—Experts agree that these turbines can be guaranteed to last 20 years. Some last longer. For example, wind turbines from the 1970s and 1980s in California, Texas, and Denmark are still in use today. Again, diligent maintenance is required for proper functioning year after year.
- **Large commercial wind turbines**—Manufacturers claim these turbines have a longevity of 20 to 25 years. Time will tell, because they have not been in use that long. As with other turbines and large machinery, repairs and regular replacement of parts are required for the equipment to last its intended lifespan. Turbines require attention, just as a car needs regular oil changes, new parts, and tune-ups.

Capital Start-Up Costs

The start-up costs cover four general areas: The cost of the wind turbine, its foundation, installation, and, for larger turbines and wind parks, access roads. Know that the wind turbine itself covers about 80 percent of capital start-up costs. The range of turbine prices varies greatly. As a rule, the bigger the turbine, the higher the cost. (Note that small standalone turbines that use batteries cost more per kilowatt than larger turbines.) Small wind turbines are not nearly as cost effective as large turbines. Improvements in the production process have lowered the costs today. For example, 25 years ago small turbines within US wind farms cost over $5,000/kW of rated power. As of 2006, that figure dropped to under $1,000/kW.

As stated, the turbine's foundation carries its own cost separate from the turbine itself **TABLE 9-1**. This cost covers not only the physical base but also the labor involved. Generally, the more experienced the construction firm, the more reasonably priced its services. The type of terrain plays a role as well.

TABLE 9-1 THE RANGE OF FOUNDATION COSTS IN THE UNITED STATES (2006)

Wind turbine	Foundation	Costs in $US	
Small wind turbine	Standard foundation	4,550	(60 $US/kW)
15.6 m rotor diameter	Excavation work	450	
75 kW (pitch controlled)	Concrete	2,300	
	Steel	1,800	
Medium-sized wind turbine	Standard foundation	27,600	(55 $US/kW)
40 m rotor diameter	Excavation work	600	
500 kW (stall controlled)	Concrete	14,000	
	Steel	13,000	
	Foundation with piles	37,000	(74 $US/kW)
	Foundation base	26,000	
	8 piles, 14 m each	11,000	
Large wind turbine	Standard foundation	61,000	(37 $US/kW)
66 m rotor diameter	Foundation with piles	83,000	(50 $US/kW)
1650 kW (pitch controlled)	(20 m)		
Large wind turbine	Standard foundation	75,000	(50 $US/kW)
64 m rotor diameter	Foundation with piles	100,000	(67 $US/kW)
1500 kW (stall controlled)	(20 m)		

Data from Hau, E. (2006). *Wind Turbines.* Berlin and Heidelberg: Springer-Verlag.

Next, you need to consider installation costs. The costs to install a turbine and then connect to the grid need to be broken down. To begin with, let's look at general installation costs for the various types of turbines:

- **Micro**—$1,500 to $2,500/m^2
- **Mini**—$1,250 to $2,500/m^2
- **Household**—$1,250 to $2,250/m^2
- **Small commercial**—$800 to $1,250/m^2
- **Large in small clusters**—$700 to $1,000/m^2

Establishing the connection between the turbine and the grid is included in the installation price **TABLE 9-2**. These costs are variable and dependent on two factors. The first is how far the turbine is to the connecting point. The second is the grid's voltage level. Large wind farms that have more than 15 MW of power demand a substation to handle the high voltage, which obviously increases costs.

Lastly, the costs for roads to provide access to the turbine must be considered as well. In fact, up to 5 percent of the price of turbines covers this kind of expense. Obviously, this applies only to medium-sized and large commercial turbines.

TABLE 9-2 EXAMPLE OF THE GRID CONNECTION COSTS FOR ONE TURBINE (2006)	Costs in $US
Wind turbine	
Rotor diameter 40 m, rated power 500 kW, induction generator coupled directly to the grid, stall control, operating voltage 400 volts	
Ex-works price	400,000
Grid connection	
Distance to connecting point 200 m, (feeder line)	
Medium-voltage grid 20 kV	
Electrical infrastructure	
■ Transformer housing, including transportation	9,000
■ Low-voltage switchboard with safety isolators, transducer, and meters	6,000
■ 0.4/20 kV transformer 500 kVA	8,000
■ 20 kV medium-voltage switchgear in SF6–design with two sections, safety circuit breaker for transformer, circuit breaker with grounding for cable outlet	7,000
■ Cable connections, electrical material, and assembly	2,500
■ Connection of the feeder line	3,500
Total of turbine-related costs	36,000
Distance-related costs	
■ Feeder line, 20 kV ground cable (200 m at 55 $US/m)	10,000
Grid connection and feeder line	46,000
Specific costs (total)	92 $US/kW
Specific costs (without feeder line)	72 $US/kW
Percentage of the ex-works price of the wind turbine	9.0%

Data from Hau, E. (2006). *Wind Turbines*. Berlin and Heidelberg: Springer-Verlag.

Financing Costs

Financing costs, or the costs incurred to pay for the investment, often come from the interest rates on bank loans. Some European banks provide special reduced interest rates for loans intended for renewable energy projects. For instance, in Germany interest rates for such projects are about 1.5 to 2 percentage points lower than standard loans. What's more, other subsidy programs can offset even these reduced interest rates; you'll read more about this later in this chapter. Something else that is gaining in popularity is project financing by banks, which involves a commitment from the borrower to pay back the loan with capital gained from the project.

Operation and Maintenance

As stated, operation and maintenance cannot be overlooked. Examples of costs that fall under this category include regular checkups, repairs, and insurance. The specific costs vary widely depending on the size and purpose of the turbine.

Needless to say, the amount of repairs required on turbines today is substantially less than what it was 20 years ago.

Just like most mechanical products on the market today, turbines come with a manufacturer's warranty. Usually, the warranty runs for a period of two years. In addition to the warranty, most manufacturers also offer a contract that outlines particular services. Contracts can offer as little as routine maintenance or as much as waiving repair costs for up to five years. Take note that the purchaser must pay for the contract, similar to an insurance policy. In 2000, contracts for smaller wind turbines of 200 to 300 kW cost between $2,000 and $4,000 per year. In contrast, contracts for 1,500 kW turbines could cost up to $15,000 per year. Most contracts do not cover the costs of operating materials or additional mechanical parts above average size. Therefore, owners must attribute about 1 percent of their total costs to repairs not covered under contract or warranty.

Additionally, purchasers have the option of three types of insurance:

- **Liability**—Covers damage to persons and property related to turbine operation. In 2006, a policy cost an average of $125/year.
- **Machine breakage**—Policies are not all-inclusive and always contain restrictions. Such policies constitute about 0.5 percent of total costs for the purchaser.
- **Loss of profit**—This applies to commercially operated wind farms. It also makes up about 0.5 percent of the purchaser's total costs.

Other operational costs exist such as land leases, taxes, and administration. Land lease prices are specific to country and location. Some landowners take an average of 5 percent of the operator's annual earnings. Fewer property owners insist on fixed lease payments. But again, this is dependent on location and the potential viability of the turbine(s). Another hidden cost is the money often paid to local communities for the disruption caused by construction and installation. Taxes are specific to the area and administrative costs usually run about 0.5 to 1.0 percent of total investment costs.

Economic Analysis

In order to determine the economic viability of a single turbine or a wind farm, various factors must be examined and analyzed from an economic perspective. To get a clear picture of costs, you must look at determining the repayment period of the initial capital investment. You must prioritize other costs and wind characteristics as well. Lastly, you must factor in the money saved from taking advantage of government-based incentives.

Simple Payback Period Analysis

One aspect of economic analysis of wind power costs involves determining the payback period of the invested capital. As a general rule, small turbines for noncommercial use should have a payback period of within 10 years. In contrast, wind farms made up of large to very large turbines can have payback periods of more than 20 years.

To determine the rate at which the capital invested must be paid back, the amount of interest is combined with the repayment. This is known as annuity FIGURE 9-3. It indicates how much capital needs to be generated to cover costs. Annuity is determined using the following formula:

$$\text{Annuity(A)} = p + \frac{p}{(1+p)^n - 1}$$

The rate of interest is represented by p and the period of capital repayment in years is n.

Cost of Energy Analysis

Answering the question of how much does wind energy cost is not a simple task. As this chapter has shown, a variety of factors must be considered to get the full economic picture. The following outlines the key factors involved in a cost of energy analysis, in order of importance:

1. Mean annual wind speed
2. The turbine's power curve
3. Investment costs
4. Strong technical availability
5. Operation and maintenance costs

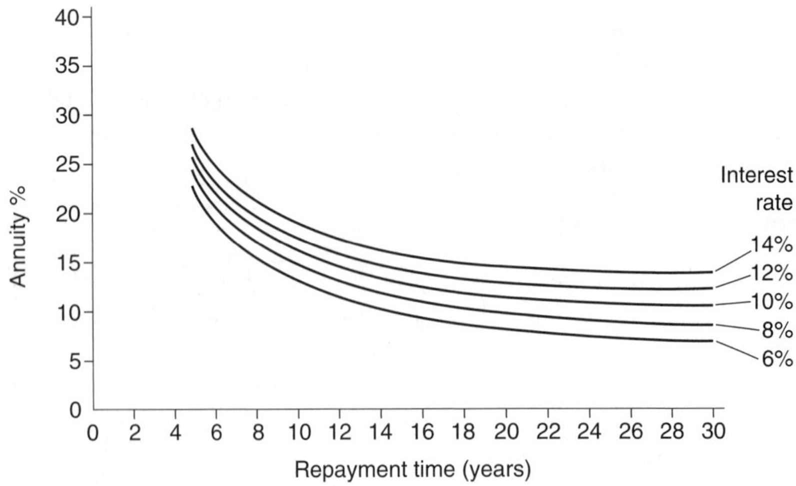

FIGURE 9-3 This graph illustrates how annuity is calculated from the interest rate and the amount of time in years in which the capital is to be repaid.

Adapted from Hau, E. (2006). *Wind Turbines*. Berlin and Heidelberg: Springer-Verlag.

Remember that repayment falls under maintenance costs. As well, the current economic climate and the predicted economic situation should be considered. Take a look at the following two tables that provide you with two typical examples. The first, TABLE 9-3, illustrates the analysis of a medium-sized wind turbine with a rotor diameter of 40 meters, 500 kW rated power, and a mean wind speed of 5.5 m/s at a hub height of 30 meters.

It is important to note that the specific investment costs with regard to annual energy yield are a telling factor in analysis. If this figure is substantially higher than $0.5/kWh, it will prove not to be economically viable.

The second example, TABLE 9-4, shows the analysis of a large wind turbine with a rotor diameter of 70 meters and 1,500 kW rated power. Its mean wind speed is 5.5 m/s at a hub height of 30 meters. Note that both of these examples provide the ideal wind speed and hub height of a good site.

TABLE 9-3 THE COSTS OF GENERATING ELECTRICITY VIA A MEDIUM-SIZED WIND TURBINE

	Costs in $US
Investment costs	
Ex-works price of the wind turbine	400,000
Planning, installation, infrastructure, and financing (30% of ex-works price)	120,000
Total investment costs	520,000
Annual costs	
Maintenance and repairs, insurance, land lease (4% of ex-works price)	16,000
Service of capital (6% interest rate p.a.), repayment period within:	
▪ 10 years (average annuity = 13.59%)	70,668
▪ 20 years (average annuity = 8.72%)	45,344
Site	
Mean annual wind speed at 30 m height	5.50 m/s
Roughness length z_0	0.1 m
Mean wind speed at hub height (50 m)	6.00 m/s
Annual energy yield	
at 98% technical availability	950,000 kWh
Specific investment costs with respect to annual energy yield	0.55 $US/kWh
Power-generation costs	
▪ with repayment in 10 years	0.091 $US/kWh
▪ with repayment in 20 years	0.065 $US/kWh

Data from Hau, E. (2006). *Wind Turbines*. Berlin and Heidelberg: Springer-Verlag.

TABLE 9-4 THE COSTS OF GENERATING ELECTRICITY VIA A LARGE WIND TURBINE

	Costs in $US
Investment costs	
Ex-works price of the wind turbine	1,400,000
Planning, installation, infrastructure, and financing (30% of ex-works price)	420,000
Total investment costs	1,820,000
Annual costs	
Maintenance and repairs, insurance, land lease (4% of ex-works price)	56,000
Service of capital (6% interest rate p.a.), repayment period within:	
■ 10 years (average annuity = 13.59%)	247,738
■ 20 years (average annuity = 8.72%)	158,704
Site	
Mean annual wind speed at 30 m height	5.50 m/s
Roughness length z_0	0.1 m
Mean wind speed at hub height (80 m)	6.50 m/s
Annual energy yield	
at 98% technical availability	3,500,000 kWh
Specific investment costs with respect to annual energy yield	0.52 $US/kWh
Power-generation costs	
■ with repayment in 10 years	0.087 $US/kWh
■ with repayment in 20 years	0.061 $US/kWh

Data from Hau, E. (2006). *Wind Turbines*. Berlin and Heidelberg: Springer-Verlag.

Incentives for Wind Energy Development in the United States

A variety of incentive programs exist today in the United States. As of 2010, the current types of subsidies and grants available to encourage wind energy development in the United States include:

- **Corporate depreciation**—Small wind turbines of 100 kW or less are eligible for credits based on decreased values of equipment.
- **Corporate tax credit**—Small wind turbines of 100 kW or less are eligible for the Business Energy Investment Tax Credit (ITC), which can cover up to 30 percent of expenses. Larger commercial turbines can make use of the Renewable Electricity Production Tax Credit (PTC) and receive 2.2 cents/kWh.
- **Federal grant programs**—One example is the Rural Energy for America Program (REAP), which provides grants for rural businesses of up to 25 percent of the wind energy project.

URBAN WIND TURBINE: FACT OR FICTION?

If more people installed wind turbines on the roofs of their homes, residents could substantially save on their electric bills. Right? Not necessarily. Due to the high amount of surface roughness in most built-up areas, turbulence is often high, so installing turbines on residential rooftops would simply not be effective in many situations. Still, this doesn't stop manufacturers from advertising rooftop urban turbines as if they were as simple for a single-family homeowner to install as energy-efficient windows. Often such ads are filled with ridiculous claims of noiselessness and high efficiency in low winds. Buyers beware!

But that doesn't mean wind power can be harnessed only on offshore wind farms or on lonely stretches of the Great Plains. Within the city limits of many big urban centers, there are areas that have low turbulence and higher winds. Examples include along riverbanks or near lakesides, in parks, and near peripheral areas of cities and towns **FIGURE 9-4**. For example, the very first urban turbine was erected in downtown Toronto in 2006 called Windshare. A group of 450 families in the area own shares in this urban turbine.

FIGURE 9-4 Wind turbines can also be found in urban areas, along riverbanks, near lakes, and within parks.
© Henryk Sadura/ShutterStock, Inc.

Some turbines are designed specifically to be located atop high-rise office or apartment towers. Such turbines are often highly efficient, but a bit costly to maintain.

In the United States, Iowa sets an example by using urban turbines in towns on the property of state-run schools **FIGURE 9-5**.

Urban turbines in Europe are nothing new. Urban dwellers have been investing in wind technology for more than 20 years. Britain is breaking new ground in this realm. The Strata Tower **FIGURE 9-6** is a 43-story residential building in London that has three built-in wind turbines at the top of the structure. These three turbines cover about 8 percent of the building's energy needs.

(Continues)

URBAN WIND TURBINE: FACT OR FICTION? (Continued)

FIGURE 9-5 A wind turbine on the property of a high school.

© Phoenix79/ShutterStock, Inc.

FIGURE 9-6 The Strata Tower in London illustrates the potential of urban wind turbine technology.

© Paul Glendell/Alamy Images

- **Federal loan programs**—USDA's Rural Energy for America Program (REAP) provides loan guarantees for up to 25 percent of a program's cost. Maximum loan guarantees are for $25 million.
- **Performance-based incentives**—The Renewable Energy Production Incentive (REPI) will pay 2.1 cents/kWh for generated electricity.
- **Personal tax credit**—The Residential Renewable Energy Tax Credit program gives up to $4,000 for wind turbines installed in 2008. Wind turbines installed after 2008 have no maximum amount of credit the owner can receive.

Chapter Summary

Determining wind turbine costs involves a variety of factors and considerations. The cost of the actual turbine covers a good portion of the total expenses, but other details such as installation, operation and maintenance, and subsidies also need to be factored into the equation. Analyzing the economic viability of a turbine or wind farm involves examination of the payback period and knowledge of what factors are of the greatest importance.

Key Concepts and Terms

Induction generator (asynchronous generator)
Net metering

Chapter Assessment: Wind Turbine Costs and Economic Analysis

1. As of 2010, all US states use net metering.
   - ❑ A. True
   - ❑ B. False
2. What is the lifespan of small commercial turbines?
   - ❑ A. 10 years
   - ❑ B. 15 years
   - ❑ C. 20 years
   - ❑ D. 25 years
3. Small turbines not for commercial use should not have a payback period of more than ________ years.
   - ❑ A. 2
   - ❑ B. 5
   - ❑ C. 8
   - ❑ D. 10
   - ❑ E. 12
4. The popularity of wind turbines and oil prices has an inverse relationship.
   - ❑ A. True
   - ❑ B. False
5. Manufacturers of micro and mini turbines state they have a maximum lifespan of ________ years.
   - ❑ A. 3
   - ❑ B. 5
   - ❑ C. 10
   - ❑ D. 12

6. Wind turbines connected to the grid usually use synchronous generators.
   - ❑ A. True
   - ❑ B. False
7. Identify the types of incentives currently in use today in the United States. (Select three.)
   - ❑ A. Corporate liability
   - ❑ B. Corporate depreciation
   - ❑ C. Corporate investments
   - ❑ D. Federal loans and grants
   - ❑ E. Personal tax credits
8. The most important factor in cost of energy analysis is:
   - ❑ A. Operation and maintenance costs
   - ❑ B. The turbine's power curve
   - ❑ C. Investment costs
   - ❑ D. Mean annual wind speed
9. Which type(s) of turbine insurance cover(s) damage to persons and property related to turbine operation?
   - ❑ A. Liability
   - ❑ B. Loss of profit
   - ❑ C. Machine breakage
   - ❑ D. A and C only
   - ❑ E. None of the above
10. The cost of a turbine's foundation is separate from the cost of the actual turbine.
    - ❑ A. True
    - ❑ B. False
11. The payback period for very large turbines is usually at least ________ years.
    - ❑ A. 10
    - ❑ B. 15
    - ❑ C. 20
    - ❑ D. 25

chapter 10

Impact on the Environment

AS YOU DISCOVER HOW wind development impacts the environment, keep in mind that any form of technology will make some sort of impact on the environment. The following chapter examines the effects of wind technology on birds and bats, the visual impact and noise created by turbines, the role of electromagnetic interference, and the primary safety issues to be considered.

Chapter Topics

This chapter covers the following topics and concepts:

- The impact wind turbines have on birds and bats and ways to minimize this impact
- Wind turbines' aesthetic impact on the environment
- Noise pollution caused by wind turbines
- Electromagnetic interference caused by wind turbines
- Safety issues related to wind turbines

Chapter Goals

When you complete this chapter, you will be able to:

- Describe the impact wind turbines have on bird and bat populations and outline the ways this impact can be minimized
- Explain how wind turbines affect the aesthetics of the environment
- Outline the various ways wind turbines generate noise pollution
- Describe how electromagnetic interference from wind turbines affects other electronic devices
- Outline the primary safety issues related to wind turbines

Birds and Bats

Many environmentalists and conservation groups do not support wind turbine technology because of the effects on birds and bats, namely death through collision. In comparison with research into other effects of wind turbines, such as noise and casting of shadows, substantially less research has gone into investigating the impact of turbines on birds and bats. For example, only 12 studies regarding turbines' impact on bat fatalities have been conducted in the United States. It will take several years more for researchers to truly understand the long-term effects of turbines on avian populations.

However, certain effects have been documented. Some studies have shown that birds indigenous to the region where wind turbines exist tend to recognize turbines as obstacles and fly around them. The situation differs for birds and bats that migrate and are not familiar with the area. Current research suggests that it is these birds and bats that have the most collisions with turbines.

Effect on Population

When bird mortality caused by turbines is compared with other bird fatalities, it is actually quite low **TABLE 10-1**. In 2006, researchers calculated that bird mortality caused by turbines was three in 100,000 deaths. In fact, the greatest risk for birds and bats is collision with other manmade structures. To put it in perspective, consider that the American Wind Energy Association (AWEA) reports that about 108,000 birds and bats lose their lives due to wind technology as compared with the 550 million that are killed by colliding into buildings and windows.

Researchers have discovered certain areas have an increased risk for avian deaths. For example, one of the earliest wind farms in the US, the Altamont Pass wind farm in California, has shown a much higher incidence of bird deaths,

TABLE 10-1 AVIAN FATALITIES CAUSED BY WIND TECHNOLOGY

Number of fatalities*	Cause
5,500	buildings/windows
1,000	domestic cats
1,000	other
800	high tension lines
700	vehicles
700	pesticides
250	communication towers
<1	wind turbines

* Number per 10,000 fatalities

Data from Erickson et al., 2002. Summary of Anthropogenic Causes of Bird Mortality.

raptors in particular. Annual deaths of birds in this area range between 7,300 and 9,600. Reasons for this elevated death rate include the facts that the population of raptors in this area is naturally high and that much of wind technology in use at the site is dated. The raptor species killed at this site include the golden eagle, red-tailed hawk, and burrowing owl.

A well-cited study by Erickson et al. (2003) identified certain areas and particular times of year when the bats are most likely to collide with turbines. The states that reported the most bat deaths caused by wind development were Colorado, Minnesota, Oregon, Wisconsin, and Wyoming. Although not all reports noted the time of year, based on more than 500 turbine farms, 90 percent of bat fatalities occurred from the middle of summer to the beginning of autumn (mid-July to mid-September). Since this study in 2003, high numbers of bat deaths occurred in Pennsylvania, West Virginia, and New York. The bats that have the highest mortality rate related to turbines are the hoary bat, the red bat, and the silver-haired bat. As of 2007, in the United States, bat mortality ranged from two to 50 bats per turbine every year.

NOTE

Upon discovering that bat deaths were much higher than thought in 2003, a partnership was formed between animal rights activists and wind energy advocates. They formed the Bats & Wind Energy Cooperative (BWEC), which is committed to reducing bat mortality through research.

Migratory Bird Treaty Act

The Migratory Bird Treaty Act (MBTA) is a convention that was put into place in 1918 to protect migrating birds, primarily from hunting. Early 20th century hunting killed many birds and threatened certain species. Often birds were killed for their feathers for women's hats and pen quills. Initially, the MBTA protected birds that migrated between the US and Canada. In 1936, migrating birds between the US and Mexico were included. Later in the 20th century, other countries made the commitment to protect migrating birds, including Japan in 1972 and the former Soviet Union in 1976.

The MBTA makes it illegal to "pursue, hunt, shoot, wound, kill, trap, capture, [or] collect" migratory birds. The law also includes their eggs, feathers, and nests. More than 800 species of migratory birds are protected under this act. Despite the fact that wind technology is not malicious and does not intentionally kill migrating birds, bird deaths by turbines are illegal under the MBTA. Enforcement of the MBTA, however, has not been consistent. Wind development projects must do their best to minimize risk to birds so as not to be prosecuted under the MBTA.

Ways to Minimize the Issue

The impact of turbines on birds and bats is relatively small, but there are various strategies to reduce it even further. The following subsections outline the six primary ways the wind industry can reduce avian deaths.

Avoiding Migration Paths

Much research goes into identifying bird migration paths. New wind development projects collaborate with wildlife organizations to ensure that wind turbines are not erected in the path of migrating birds. Daily and long-term paths are examined and avoided.

Fewer and Larger Turbines

Reducing the number of turbines by increasing turbine size is advantageous to avian life. Larger turbines have a lower frequency rotor rotation. Birds are more likely to notice the lower rotational speed of the blades, which allows them to fly around the turbine.

Site Selection

Much assessment is required to determine if a particular site poses a high risk for avian populations. Care is taken to erect turbines and build roads that do not break up known habitat zones. Additionally, developers must investigate if common birds of prey reside near the potential site.

Nest Removal

Birds sometimes nest in and around turbines, depending on the local environment, and such nests can't always simply be removed. If the birds that made them are of a protected species, developers need to check with local authorities on how to remove a nest legally.

Burying Electrical Lines

Any electrical or power lines should be buried. Foraging and burrowing birds are at risk for electrocution if these lines are left exposed.

Alternate Tower Designs

It is now known that towers that use guyed wires present high risk to birds and bats, so their use is not recommended. Research also shows that steady lights attract some species of birds and bats. Today, regulations require that flashing red lights be used in their place. Some wind developers use ultrasonic sounds to deter certain species of bats from a potential site. Painting blades in banded colors has also been tried, albeit with little success.

Additionally, an alternative form of rotor is available that seems to reduce bird fatalities. Instead of the traditional bladed design, this rotor has a tulip shape that seems to be easier for birds to see. It gives the impression of a solid, static object like a building, so birds and bats know to avoid it. The WindTulip is one such design, as are compressed field VAWTs.

Aesthetic Impact on the Environment

As wind energy becomes a viable solution for the future, more wind farms will be constructed. This means that turbines will also come under fire if they damage the environment. Most people agree that wind turbines impact the visual landscape. What can't be agreed upon is how much of an impact turbines make **FIGURE 10-1**. Despite the subjective interpretation of aesthetics, Swedish researchers assert that the amount of aesthetic impact depends on the following three elements:

- **Psychological factors**—What connections do people make with turbines; or, in other words, what do people think of when they see a turbine? This can happen consciously or unconsciously.
- **Topography**—Is the site of the turbine open or closed? A wind turbine in a field affects the visual landscape much differently than a wind turbine in an urban area or a forest.
- **Wind turbine height**—How tall is the turbine? Generally speaking, a turbine that stands 50 m or less does not greatly disturb the visual landscape, especially if it is placed in environments with other structures, man-made or otherwise. A height of 50 m or more is usually visible at far greater distances and thus disturbs the visual landscape.

Wind Turbine Visibility and Appearance

One concern specific to the visual landscape is shadow effects caused by turbines. Shadows are a nuisance for people inside buildings that are in close proximity to turbines. Although other man-made structures cast shadows, shadows from

FIGURE 10-1 Most people agree that wind turbines impact the visual landscape. What cannot be agreed upon is how much of an impact turbines make.

turbines are different because they are moving objects. What results from rotating blades is **shadow flicker.** For some people, adverse physiological effects are possible. Additionally, glare off rotating blades upsets those in the immediate area.

Mitigating the Problem

In addition to considering the three factors outlined by Swedish researchers, other factors can minimize the visual impact of turbines. They include:

- Color—It is recommended that turbines adhere to white, off-white, light gray, or beige. Color should be uniform and without designs or advertisements.
- Spacing—Some experts believe that a balanced configuration is more pleasing to the eye. Erecting turbines on a flat terrain in rows is preferred.
- Blades—Three-blade turbines are recommended. On wind farms, the blades of the turbines should all rotate in the same direction. Blades should be covered with a non-reflective substance to diminish glare.
- Number of turbines—Ideally, having fewer but larger turbines makes less of an impact than having many smaller ones.

Noise Pollution

A great deal of research has gone into investigating how wind turbine noise impacts the environment. Not all turbines create the same levels of noise. As well, turbines generate different types of noise. Before exploring the main types of noise emissions, know that country-specific regulations exist for acceptable noise levels

TABLE 10-2 ACCEPTABLE NOISE LEVELS IN GERMANY, THE NETHERLANDS, AND DENMARK

Permissible noise level (dB(A))	Germany	Netherlands	Denmark
Predominantly industrial area			
day:	65		
night:	50		
Mixed industrial and residential area			
day:	60	50	
night:	45	40	
Predominantly residential area			
day:	55		
night:	40		
Exclusively residential area			
day:	50	45	
night:	35	35	40
Rural area			
day:	40		
night:	30	45	

Data from Hau, E. (2006). *Wind Turbines*. Berlin and Heidelberg: Springer-Verlag.

TABLE 10-2. The location of the turbine and the time of day are the two determining factors. Note that the measurement used to describe the noise level is the sound pressure level, represented by dB(A).

Different size turbines generate varying noise levels. TABLE 10-3 outlines the ideal sound pressure levels for turbines based on size.

Additionally, it is important to know that noise that naturally occurs in the environment is intensified with wind speed. This means that at certain wind speeds turbine noise is not noticed due to the existing noise in the environment TABLE 10-4. Generally, if environmental noise is 6 dB(A) greater than turbine noise, turbine noise will be not detected. As a rule, wind-generated background noise will increase 2.5 dB(A) per 1 m/s of wind speed.

TABLE 10-3 IDEAL SOUND PRESSURE LEVELS AS A FUNCTION OF SIZE

Turbine size	Rotor diameter	Ideal sound pressure levels dB(A)
Small	Up to 20 m/100 kW	~95
Medium	Up to 40 m/500 kW	~98
Large	70-80 m/2000 kW	103–105
Multi Megawatt	100-120 m	105–107

TABLE 10-4 COMPARISON OF WIND FARM NOISE AND COMMON TYPES OF AMBIENT NOISE

Source/Activity	Indicative noise level (dB)
Threshold of hearing	0
Rural nighttime background	20–40
Quiet bedroom	35
Wind farm at 350 m	35–45
Busy road at 5 km	35–45
Car at 65 km/h at 100 m	55
Busy general office	60
Conversation	60
Truck at 50 km/h at 100 m	65
City traffic	90
Pneumatic drill at 7 m	95
Jet aircraft at 250 m	105
Threshold of pain	140

Data from CIEMAT

Infrasound

Infrasound is defined as low frequency noise that registers below 20 Hz. Turbines are believed to generate this low frequency noise. This type of noise can be audible or inaudible, but generally humans cannot detect infrasound. At present, there is no agreement among scientists and wind development experts whether low frequency sound caused by turbines is a health threat. If you search online for "infrasound and wind turbines," most of the information you will find discusses the adverse physical effects of this low frequency noise. Common symptoms include headaches, irritability, sleep difficulties, ringing in the ears, and dizziness. Low frequency sound has been known to trigger other reactions in addition to physical side effects.

Physical Sensations

In the range of 4 to 25 Hz, some people feel changes in their body. Vibrations can be felt from the waist down, as well as a sensation of pressure in the throat and upper chest region. Another consequence of infrasound caused by wind turbines is a feeling of static pressure.

Rattling of Doors and Windows

One last additional effect of wind turbine infrasound is the rattling of doors and windows in structures near the turbines. Closing windows or doors to minimize the shaking does not work. In fact, infrasound and its effects intensify when contained, so leaving windows open reduces rattling. Low frequency noise of less than 20 Hz has been documented to shake the structural frames of homes.

Aerodynamic Blade Noise

The main culprit for wind turbine noise comes from the air that flows around the rotor, referred to as aerodynamic blade noise. Most of the noise generated from the rotor originates from the outside portion of the blade. Wind developers have discovered that if the blade-tip speed is reduced by 25 percent, aerodynamic blade noise can be reduced by approximately 6 dB(A). Reducing a turbine's power output can also diminish blade noise but the most important factor is the blade-tip speed, especially for rotational speeds under 65 m/s. As well, incorporating the use of pitch control can help reduce aerodynamic noise.

Mechanical Noise

The last type of wind turbine noise is mechanical. The gearbox, hydraulic pumps, gear motors, and the generator's cooling system can generate mechanical noise in wind turbines. Much attention has been given to the gearbox, considered the most important factor related to mechanical noise. Keep in mind that a silent gearbox, to date, does not exist. Most commonly, noise from the gearbox travels through the nacelle and the tower. Towers that are hollow and made of steel cause the noise to intensify, making soundproofing gearboxes a necessity.

> **NOTE**
>
> The noise emitted from a wind farm is the combined noise of the individual wind turbines.

Electromagnetic Interference

Electromagnetic interference (EMI) refers to disturbance and/or damage caused by other electronic devices. Wind turbines have been known to cause EMI and affect common electronic devices. The extent of EMI effects is related to the distance of the turbines from the electronic device, the type of rotor blade design, and topography. Research has shown that rotor blades consisting of steel trigger the highest levels of interference. Today, blades are made of composite materials. Worth noting however, is that blades are treated with lightning protection, which can result in higher levels of EMI.

The following seven sub-sections list the most commonly affected devices.

Television

Wind turbines are known to interfere with television signals. The primary cause is the reflection of television signals off a turbine's blades, which disperses the signals **FIGURE 10-2**. Today, often readjusting television antennas is usually sufficient to correct the problem. When that is not effective, a small transmitter can be used to deflect the EMI from the turbines.

Radio

The EMI effects from wind turbines on both AM and FM radio are minimal. Furthermore, turbines can affect radio signals only when in close proximity. Literature cites less than 100 meters. AM radio signals usually have signal wavelengths longer than 100 meters so EMI is unlikely. In regard to FM radio, research has shown that EMI from nearby turbines affect the radio signal only if it lacks in strength.

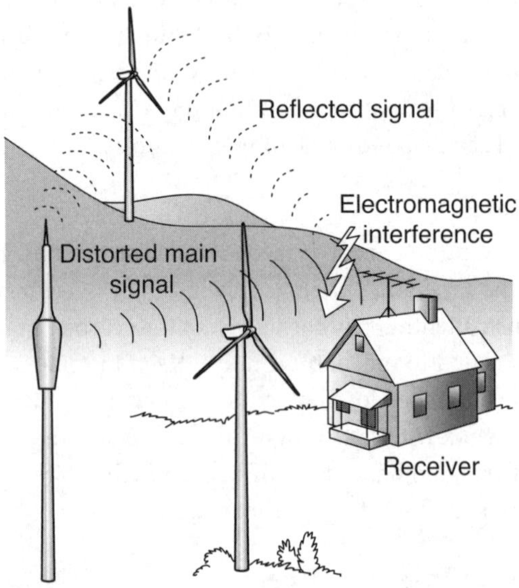

FIGURE 10-2 This illustration shows how television signals can be reflected and scattered by a turbine's electromagnetic energy.

Adapted from Spring-Verlag Gmbh and Co.

Aviation Navigation and Landing Systems

EMI occurs within a small area, which is advantageous to aviation navigation and landing systems because they do not rely on large areas of reception. Because the area of EMI is contained, wind developers can choose to erect wind turbines outside locales with such navigation systems.

Radar

To date, the effects of turbine EMI on radar, such as radar used by meteorologists or the US Department of Defense, are not well understood. Additional investigation is required so that certification standards and regulations can be put into place to minimize turbine EMI on radar.

Microwave Links

Microwave links react to EMI from turbines in the same way as fixed radio links. Note that fixed radio links always have a specific receiver that picks up the signal. In order for the transmission to work properly a clear pathway between the transmitter and the receiver is necessary. If EMI from a turbine crosses this path, the transmission will be weakened or disrupted altogether.

Cell Phones

The strength of a cell phone signal can be affected by the topography of an area and the location of the receiver. Often, a slight shift in the position of the receiver can improve a weak signal and eliminate EMI from nearby turbines.

Satellite Services

Satellite services can be disrupted in the same way microwave and radio links can be disturbed by wind turbine EMI. Wind developers need to be aware of satellite paths before building wind farms to ensure signals will not be disturbed.

Safety Issues

Before exploring the safety issues related to wind technology, keep in mind other forms of technology, such as automobiles and air travel, are much more dangerous than wind turbines. That said, safety concerns exist, so they must be mentioned.

Blade Throw

Of great concern is the damage that can be done by broken pieces of a turbine's rotor. The most common cause of a broken blade can traced to the rotor-brake system. When the brakes fail to work, the rotor speed increases to a level that is called rotor runaway. With no control over speed, the blade can fracture or

FIGURE 10-3 A three-blade wind turbine with a collapsed blade.

collapse **FIGURE 10-3**. Each turbine is different, but for large turbines, when the rotor speed is double or triple the rated rotor speed, blade damage occurs. Although blade throw is rare, it did happen with older turbines in the past. Today's turbines have at least two separate rotor-braking systems that prevent blade damage, and thus blade throw.

Take note that blade fracture can also occur from hidden stress in the blade. This means that a blade can break even when the rotor is functioning at normal speeds. There is one documented case of this occurring. One of the blades of an older large turbine (Smith-Putnam) had a weakness at its base. The 8-ton blade broke away from the turbine and landed about 230 m away.

Because it is difficult to accurately predict the distance a broken blade may be thrown, developers should take prudent safety precautions. It is recommended that wind turbines be surrounded by a buffer zone between the turbines and populated areas.

Falling Ice

When a news source reports that large chunks of ice, three to four feet long, have been thrown from spinning turbine blades, people look at wind technology with skepticism. Although ice can fall from turbines, just as ice can accumulate and drop from buildings and trees, this is rare. Furthermore, wind turbine technology today takes into account a buildup of ice on rotor blades. Turbines have an automatic mechanism that stops the blades from rotating when too much ice builds up.

Naturally, if a turbine or wind farm is in close proximity to populated areas, the possibility of falling ice must be considered. For instance, studies have been done in the Swiss Alps close to ski resorts and they indicate that risk exists. The greatest risk of harm from falling ice was directly below the turbine, but 5 percent of the ice fell more than 80 m from the base FIGURE 10-4.

Other studies confirm that ice throw is a possibility. Wind developers must include falling or thrown ice into their site assessment so that it does not pose a safety hazard.

Tower Failure

Wind turbine towers in North America and Europe have collapsed in the past and news coverage of such events is intense. However, such occurrences are far from the norm. It is more accurate to say tower failure is highly unlikely, especially if all the established safety standards are followed during construction and maintenance.

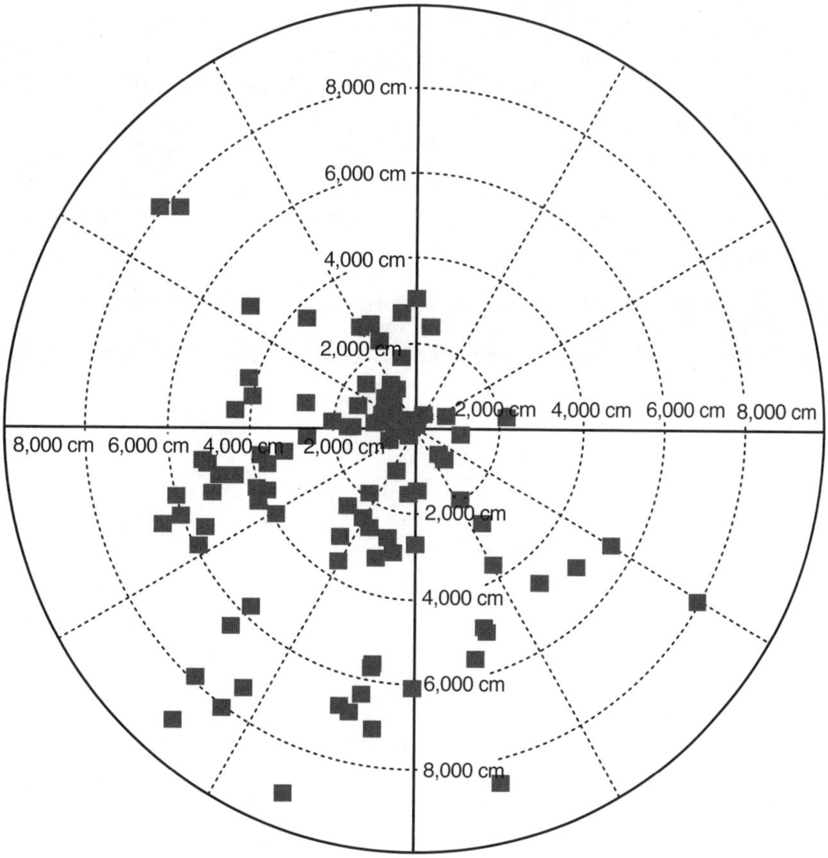

FIGURE 10-4 A diagram of the results from the Swiss Alps study, which shows where ice fell in relationship to the turbine (center).

Data from Wind Turbine Ice Throw Studies in the Swiss Alps

DO WIND FARMS AFFECT THE WEATHER?

Does that sound like a trick question? It's not. Even though wind technology is viewed as a green alternative to fossil fuels, its effects on local weather patterns and climate change are still being studied. What researchers do know is that air in the wake of the rotor is mixed in an up-and-down motion, which causes some warming and drying effects. For example, studies have shown that surface temperatures just prior to sunrise can increase by 4 degrees F, which in turn can lead to drier soil. This could be good news for crops in wetter seasons, because the turbulence could help keep plants drier, reducing the incidence of fungus and damage from moisture-loving insects. During droughts, however, the turbulence might be less welcome if the plants required more irrigation.

If overall temperatures were to increase in the hotter months, people might run air conditioners more. This rise in electricity use might even override the savings gained from wind technology.

The bigger question of whether a number of very large wind farms could impact global warming has not yet been answered. One study involved a computer-simulated model of a wind farm consisting of thousands of turbines that covered the Midwestern US. The results showed an average drop in wind speed of about six miles per hour. Additionally, storm behavior changed.

If you consider that climate change is under way, the effects of wind farms on the environment may help negate some climate changes. Since a wind turbine removes some energy from the atmosphere, there is an overall lessening of total energy, even if just by a small amount. So on a global scale, wind turbines may have a future in positively impacting some aspects of climate change.

CHAPTER SUMMARY

This chapter has provided you with a broad look at wind technology's impact on the environment. Relative to other energy-producing technologies, wind development leaves barely a footprint, but that does not stop wind specialists and environmentalists from trying to minimize its impact. These include decreasing the negative effects on avian populations, increasing awareness regarding the aesthetic impact of wind farms, reducing wind turbine noise, limiting electromagnetic interference with electronic devices, and staying tuned to safety concerns.

KEY CONCEPTS AND TERMS

Shadow flicker
Infrasound

CHAPTER ASSESSMENT: IMPACT ON THE ENVIRONMENT

1. Glare that is reflected off wind turbine blades can cause EMI with television signals.
- ❑ **A.** True
- ❑ **B.** False

2. The ________ is often the cause of mechanical turbine noise.
- ❑ **A.** generator
- ❑ **B.** gearbox
- ❑ **C.** rotor
- ❑ **D.** blades
- ❑ **E.** None of the above

3. What is the maximum amount of Hz for infrasound?
- ❑ **A.** 2
- ❑ **B.** 20
- ❑ **C.** 100
- ❑ **D.** 200
- ❑ **E.** 250

4. Aesthetically speaking, blades should rotate in the same direction.
- ❑ **A.** True
- ❑ **B.** False

5. What strategies are implemented to minimize avian death? (Select two.)
- ❑ **A.** Rerouting birds to new migratory paths
- ❑ **B.** Building smaller turbines
- ❑ **C.** Avoid the use of guyed wires
- ❑ **D.** Bury electrical lines

6. Larger wind turbines present a higher risk factor for avian populations.
   - ❑ A. True
   - ❑ B. False
7. Some studies suggest that 90 percent of bat fatalities caused by wind turbines occur between mid-July and:
   - ❑ A. mid-August.
   - ❑ B. mid-September.
   - ❑ C. mid-October.
   - ❑ D. mid-November.
   - ❑ E. None of the above
8. Which colors are recommended for wind turbines? (Select three.)
   - ❑ A. Light gray
   - ❑ B. Ivory
   - ❑ C. White
   - ❑ D. Pale yellow
   - ❑ E. Dark beige
9. FM radio signals usually have a wavelength of longer than 100 meters.
   - ❑ A. True
   - ❑ B. False
10. What is the minimum sound pressure level of environmental noise required to mask turbine noise?
    - ❑ A. 6 dB(A)
    - ❑ B. 16 dB(A)
    - ❑ C. 60 dB(A)
    - ❑ D. None of the above
11. What are the two primary causes of blade fracture?
    - ❑ A. Disguised blade stress
    - ❑ B. Gearbox malfunction
    - ❑ C. Structural weakness in tower
    - ❑ D. Rotor runaway
12. Flashing red lights on turbines have been replaced with steady white lights because the former caused increased bird collisions.
    - ❑ A. True
    - ❑ B. False
13. Which countries/regions are covered under the Migratory Bird Treaty Act? (Select three.)
    - ❑ A. US
    - ❑ B. Former Soviet republics
    - ❑ C. China
    - ❑ D. Central America
    - ❑ E. Mexico

14. Newer turbines are equipped with blades that detect heavy ice loads.
- ❑ A. True
- ❑ B. False

15. The primary cause of aerodynamic noise can be attributed to the:
- ❑ A. nacelle.
- ❑ B. generator.
- ❑ C. gearbox.
- ❑ D. rotor.

Answer Key

CHAPTER 1 History

1. B **2.** B **3.** B, D, and E **4.** A **5.** B **6.** A **7.** D **8.** B **9.** A
10. C **11.** B, D, and E **12.** A **13.** B and D **14.** C **15.** B

CHAPTER 2 Basic Concepts

1. A **2.** D **3.** B **4.** C and D **5.** B **6.** A **7.** A **8.** A and D **9.** B
10. A **11.** A **12.** A and C **13.** D **14.** C **15.** B

CHAPTER 3 Wind Energy Converters

1. B **2.** C **3.** B and D **4.** B **5.** B, C, and D **6.** A **7.** D **8.** A
9. B **10.** D

CHAPTER 4 Rotor Aerodynamics

1. B **2.** A **3.** B **4.** B **5.** A, D, and E
6. A **7.** A **8.** D **9.** B **10.** B

CHAPTER 5 Wind Resource

1. A **2.** C **3.** D **4.** B **5.** A and D **6.** D **7.** A, B, and D **8.** A
9. C **10.** B **11.** B **12.** A **13.** A **14.** B **15.** C

CHAPTER 6 Power Output and Energy Yield

1. A **2.** C **3.** C **4.** A and C **5.** B, C, and E **6.** B **7.** C
8. A **9.** A **10.** B

CHAPTER 7 Commercial Applications

1. A 2. B 3. C 4. A 5. B 6. A and D 7. A, C, and D 8. A 9. B 10. C

CHAPTER 8 Offshore Wind Energy

1. A 2. C 3. A 4. B and D 5. B 6. B 7. B 8. A and B 9. C 10. A 11. A and C 12. D

CHAPTER 9 Wind Turbine Costs and Economic Analysis

1. B 2. C 3. D 4. A 5. D 6. B 7. B, D, and E 8. D 9. A 10. A 11. C

CHAPTER 10 Impact on the Environment

1. A 2. B 3. B 4. A 5. C and D 6. B 7. B 8. A, B, and C 9. B 10. A 11. A and D 12. B 13. A, B, and E 14. A 15. D

GLOSSARY

Air density Describes the mass per unit volume of air.

Alternating current (AC) A flow of electricity that moves back and forth in two opposing directions. Most turbines produce an alternating current.

Anchor The structure found at the base of a wind tower that connects the guy wires to the tower.

Aspect ratio A parameter that indicates blade slenderness.

Average wind speed A type of wind speed measurement that provides the average of a determined number of instantaneous wind speeds.

Base The foundation of a wind tower.

Caisson A watertight container within which construction work is carried on under water.

Charge controller A piece of machinery whose primary function is to protect batteries from getting too much charge.

Concentrator A mechanism that enhances available wind.

Conduit The protective plastic or metal tubing that encases the transmission wires.

Direct current (DC) A flow of electricity that moves in one direction and remains consistent. Generators that produce a direct current are costly and require much maintenance, so they are not commonly used. The exception is very small wind turbines that recharge batteries.

Distributed generation The practice of incorporating wind power for residences and urban centers.

Diurnal variations The wind changes that occur over the period of a day.

Drag The aerodynamic force that slows down the rotational movement of the rotor.

Dutch windmill A windmill developed in the Netherlands in the 16th century that operated on a fixed base and used a rotating tower cap that turned the wind wheel.

Frost line Indicates up to what point the ground freezes in winter.

Gearbox A mechanism that works in conjunction with the generator to convert mechanical energy into electrical energy.

Generator The mechanism found at the core of the electrical system that converts mechanical energy into electrical energy.

Geostrophic wind The airflow that is found above all manmade or natural obstacles.

Guy strand An inflexible yet strong cable that contains few strands of other wires.

Guy wires The metal cables that are used to secure fixed guyed towers and tilt-up towers to the ground.

Hollow-post mill A windmill created in 15th-century Holland used to pump water. It is also referred to as a "Wipmolen" or "Kokermolen."

Horizontal axis Describes how the wind wheel of the traditional windmill in the 12th century rotated parallel to the ground.

Induction generator (asynchronous generator) A type of electrical generator in which the speed of the rotor and the speed of the magnetic field created by the stator are not the same.

Infrasound Low frequency noise.

Instantaneous wind speed A type of wind speed measurement that examines the velocity of the wind in the moment.

Inverter A mechanism that transforms direct-current (DC) electricity to alternating-current (AC) electricity.

Lift The aerodynamic force that causes the rotational movement of the rotor.

Net metering A policy that allows utility customers who produce their own electricity to sell it back to the grid and pay only the net difference between the energy they get from the grid and the energy they send back to the grid.

Peak wind speed A type of wind speed measurement that provides the highest wind speed possible on a given site.

Post windmill A windmill developed in 12th-century Europe that rotated on a horizontal axis and was used to mill grain. It is also referred to as a trestle windmill.

Power coefficient (C_p) The number that tells how well a wind turbine converts wind to electricity.

Prandtl layer The area of the boundary layer near the earth's surface.

Rated power A term used in the wind industry that reflects the peak watts in a power curve.

Rebar The metal rod that sits inside the concrete base or anchor of a tower to provide extra support.

Renewable energy source A type of natural energy, such as wind, tides, or sunlight, that can be naturally replenished.

Rotor The blades and the hub of a turbine.

Rotor diameter The longest distance between one side of the circle created by the rotor blades and the other side.

Rotor power coefficient (C_{PR}) The numeric value that illustrates how effectively a wind turbine functions.

Shadow flicker The strobe-like effect that is caused by the rotation of the turbines' blades.

Spring sail A type of wind wheel sail that took the form of wooden shutters, which were hinged to an iron rod. Invented by Andrew Meikle in 1792.

Standard power rating The measurement used to describe the amount of energy produced by the turbine.

Stator The stationary or unmoving part of the rotor.

Swept area A variable in the wind power equation that describes the area that the rotor covers.

Synoptic variations Wind changes that occur over a period of a few to several days.

Tip-speed ratio The relationship between wind speed and the rotational speed of the blade tip.

Torque A force that turns or twists.

Tower windmill A type of windmill that had its wind wheel based on a round stone tower.

Transmission wires The electrical wires that carry the electricity from the turbine to the ground so it can be put to use.

Turbulence Wind speed variations that occur within short time frames, such as minutes or seconds.

Vertical axis Describes how the wind wheel rotates on a right angle to the horizon.

Watt (W) A unit of measurement that describes the rate of energy generation, movement, or use.

Watt-hours (Wh) A unit of measurement that describes electrical units.

Yaw bearings A type of bearing in a wind turbine that allows complete revolving motion of the turbine head.

Yawing The act of rotating on a vertical axis.

Yaw system The mechanism in a wind turbine that works to ensure the rotor faces the wind.

References

American Wind Energy Association. (2009). "About small and community wind." Retrieved August 31, 2010, from http://www.smallandcommunitywindexpo.org/about.cfm

American Wind Energy Association. (2010). "About AWEA." Retrieved October 5, 2010, from http://www.awea.org/about/index.html

Aquirre, E.L. (2009.) $500K Grant to Study Cause, Effect of Manufacturing Defects. University of Massachusetts Lowell. Retrieved September 12, 2010, at http://www.uml.edu/media/enews/Niezrecki_wind_turbines.html

Basic Aerodynamic Operating Principles of Wind Turbines. (2009.) American Wind Energy Association. Retrieved June 24, 2011, from http://archive.awea.org/faq/vawt.html

Bean, P. & Kollins, K. (2007, April 24). Wind turbines and their impacts on avian and bat populations. Duke University: Nicholas School of the Environment and Earth Sciences. Retrieved November 21, 2010, from http://www.duke.edu/web/nicholas/bio217/ptb4/index.html

Blade Snaps Off Huge Wind Turbine. (2010, March 23). BBC. Retrieved September 12, 2010 from http://news.bbc.co.uk/2/hi/uk_news/scotland/glasgow_and_west/8582554.stm

Bogo, J. (2007, September 14). "How the deadliest wind farm can save the birds: Green machines." *Popular Mechanics*. Retrieved June 29, 2011, from http://www.popularmechanics.com/science/environment/green-energy/4222351

Burton, T., Sharpe, D., Jenkins, N., & Bossanyi, E. (2008). *Wind energy handbook*. West Sussex, England: John Wiley & Sons Ltd.

Cartledge, J. (2010, May 21). 3M to conduct year-long study on wind turbine blade erosion. Bright Energy. Retrieved September 12, 2010 from http://www.brighterenergy.org/10753/news/wind/3m-to-conduct-year-long-study-on-wind-turbine-blade-erosion/

"China to boost offshore wind power generation." (2011, May 16). Xinhua News Agency. Retrieved June 23, 2011, from http://www.chinadaily.com.cn/china/2011-05/16/content_12520766.htm

Chittam, G. (2010, March 23). Green energy powers coastal community services. King 5 News. Retrieved on September 24, 2010, from http://www.king5.com/news/environment/Green-Energy-Powers-Coastal-Community-Services-88960222.html

Clarke, S. (2003, September). Electricity generation using small wind turbines at your home or farm. Ontario Ministry of Agriculture, Food and Rural Affairs.

Retrieved August 16, 2010, from http://www.omafra.gov.on.ca/english/engineer/facts/03-047.htm

Community wind toolbox Chapter 1: Introduction to community wind development. (2010). Windustry. Retrieved August 30, 2010 at http://www.windustry.org/your-wind-project/community-wind/community-wind-toolbox/chapter-1-introduction/community-wind-toolbox

Community Wind Toolbox Chapter 5: Siting Guidelines. (2010). Windustry. Retrieved November 21, 2010, from http://www.windustry.org/your-wind-project/community-wind/community-wind-toolbox/chapter-5-siting-guidelines/community-wind-t

Community wind projects get a boost from Washington State. (2010, August 30). Windustry. Retrieved September 1, 2010, at http://www.windustry.org/news/community-wind-projects-get-a-boost-from-washington-state

Coppin, P.A., Ayotte, K.A., & Steggel, N. (2003). Wind Resource Assessment in Australia—A Planners Guide, v. 1.1. Canberra: Wind Energy Research Unit, CSIRO Land and Water (Australian Commonwealth Scientific and Industrial Research Organisation).

Delabole Wind Farm – New Turbines. (2011, February 22). *YouTube uploaded by http://www.goodenergy.co.uk*. Retrieved August 13, 2011, from http://www.youtube.com/watch?v=U9xZvROcAqY

Delabole Wind Farm's Tall Turbines Power More Homes. (2010, August 2). BBC News. Retrieved August 20, 2010, from http://www.bbc.co.uk/news/uk-england-cornwall-10834942

Dmitriev, G. (2001). Wind Energy in Russia. Inforse – Europe: The International Network for Sustainable Energy. Retrieved January 24, 2012, from http://www.inforse.org/europe/wind_RU_en.htm

Energy resources: Wind power. (2010, Sept 24). Energy Resources. Retrieved September 24, 2010, from http://www.darvill.clara.net/altenerg/wind.htm

Environmental impacts of wind-energy projects. (2007). The National Academies Press. Retrieved November 23, 2010, from http://www.nap.edu/openbook.php?record_id=11935&page=140

Epstein, R. (2010, February 23). What is community wind? *Energy Bulletin*. Retrieved August 31, 2010, from http://www.energybulletin.net/node/51680

European Wind Energy Association (2010). "Offshore wind heads for record year" (press release). Retrieved August 8, 2010, from http://www.ewea.org/index.php?id=60&no_cache=1&tx_ttnews[tt_news]=1865&tx_ttnews[backPid]=1&cHash=c0f89ca1cbe22ef183866d16d86d38a9

Federal Incentives/Policies for Renewables & Efficiency. (2010). DSIRE™ Database of State Incentives for Renewables & Efficiency. Retrieved November 12, 2010, from http://www.dsireusa.org/incentives/index.cfm?state=us&re=1&EE=1

Foundations for offshore turbines. (2010). Offshore-Wind.de. Retrieved October 31, 2010, from http://www.offshore-wind.de/page/index.php?id=10236&L=1

Galbraith, K. (2008, December 9). Ice-tossing turbines: Myth or hazard? *The New York Times*. Retrieved November 24, 2010, from http://green.blogs.nytimes.com/2008/12/09/ice-tossing-turbines-myth-or-hazard/

Gipe, P. (2006, May 1). Urban Turbines: What They Are and What Makes Sense. Wind-Works. Retrieved November 12, 2010, from http://www.wind-works.org/articles/UrbanWindTurbines.html

Gipe, P. (2009). *Wind energy basics*. White River Junction, VT: Chelsea Green Publishing Company.

Gipe, P. (2004). Wind power: Renewable energy for home, farm, and business. White River Junction, VT: Chelsea Green Publishing. Retrieved October 4, 2010 from http://books.google.com/books?id=1ZKky0XK6IUC&pg=PA61&lpg=PA61&dq= swept+area+method+estimating+turbine+performance&source=bl&ots=ei4FxeDReF&sig=knx1AU44vummueThMGz8sF1wMkU&hl=en&ei=Qx2qTPWNMcP48AbMn9nWDA&sa=X&oi=book_result&ct=result&resnum=1&ved=0CBcQ6AEwAA#v=onepage&q&f=false

Harding, D. (2008, April 21). What Is Net Metering? Calfinder: Nationwide Home Solar Power Contractors and Information. Retrieved November 12, 2010, from http://solar.calfinder.com/blog/solar-information/what-is-net-metering/

Hau, E. (2006). *Wind turbines*. Berlin and Heidelberg: Springer-Verlag.

Khatri, D. (2010, February 17). "Economics of taller wind towers." *Renewable Energy World North America Magazine*. Retrieved August 20, 2010, from http://www.renewableenergyworld.com/rea/news/article/2010/02/economics-of-taller-wind-towers

LaMonica, M. (2008, August 28). "GE reshapes the future of wind power." Cnet News. Retrieved August 8, 2010, from http://news.cnet.com/8301-11128_3-10023553-54.html

Layton, J. (2010). How Wind Power Works. *How Stuff Works*. Retrieved September 10, 2010 from http://science.howstuffworks.com/environmental/green-science/wind-power3.htm

Leask, D. (2010, April 18). Glue not UFO caused turbine blade failure. News Scotsman. Retrieved September 12, 2010, from http://news.scotsman.com/wind-power/Glue-not-UFO-caused-turbine.6235114.jp

Lilley, M.B. & Firestone, J. (2008). Wind power, wildlife, and the Migratory Bird Treaty Act: A way forward. BNET: The CBS Interactive Business Network. Retrieved November 16, 2010, from http://findarticles.com/p/articles/mi_hb3153/is_4_38/ai_n31182212/?tag=content;col1

Low frequency noise and wind turbines. (2005, February). British Wind Energy Association. Retrieved November 22, 2010, from http://www.bwea.com/ref/lowfrequencynoise.html

Manwell, J.F., McGowan, J.G., & Rogers, A.L. (2010). Wind energy explained: Theory, design and application. Google Books. Retrieved November 23, 2010, from http://books.google.com/books?id=roaTx_Of0vAC&dq

Marshall, B. & Adkins, B. (2010). How Airplanes Work. *How Stuff Works*. Retrieved September 10, 2010, from http://science.howstuffworks.com/transport/flight/modern/airplane.htm

McCurry, J. (2010, September). Energy Innovations. *Site Selection Online*. Retrieved November 12, 2010, from http://www.siteselection.com/features/2010/sep/Energy-Innovations/

Migratory birds and habitat programs. (n.d.). U.S. Fish & Wildlife Service. Retrieved November 22, 2010, from http://www.fws.gov/pacific/migratorybirds/mbta.htm

Minnesota hospital working to build a community wind farm. (2010, August 13). Renewable Energy World. Retrieved August 30, 20101, from http://www.renewableenergyworld.com/rea/news/article/2010/08/minnesota-hospital-working-to-build-a-community-wind-farm

Nelson, R. (2009, December 26). Where are the best locations for wind farm installations? Renewable Power News. Retrieved on September 24, 2010, from http://www.renewablepowernews.com/archives/796

New Tower Reaches High to Catch the Wind. (2009, May 15). *Renewable*

Energy World. Retrieved August 20, 2010 from http://www.renewableenergyworld.com/rea/news/article/2009/05/new-tower-reaches-high-to-catch-the-wind

New wind turbine design good for rural, urban environment. (2008, June). *YouTube uploaded by Voice of America News*. Retrieved May 20, 2011, from http://www.youtube.com/watch?v=fNudnI5tzf8&feature=player_embedded

Offshore Support Structures. (2009). *Wind Energy: The Facts*. Retrieved October 31, 2010 from http://www.wind-energy-the-facts.org/en/part-i-technology/chapter-5-offshore/wind-farm-design-offshore/offshore-support-structures.html

O'Neal, R.D., Hellweg, Jr., R.D. & Lampeter, R.M. (2009, July). A study of low frequency noise and infrasound from wind turbines. Epsilon Associates, Inc.

Onshore impacts. (n.d.). *Wind energy: The facts*. Retrieved November 22, 2010, from http://www.wind-energy-the-facts.org/en/environment/chapter-2-environmental-impacts/onshore-impacts.html

Pentland, W. (2008, July 9). America's best places for alternative energy. *Forbes*. Retrieved on September 24, 2010, from http://www.forbes.com/2008/07/09/energy-solar-green-biz-energy-cx_bp_0709atlas.html

Power Coefficient, The. (2003, June 1). The Danish Wind Industry Association. Retrieved September 12, 2010 from http://www.talentfactory.dk/en/tour/wres/cp.htm

Pros and Cons of an Offshore Wind Farm. (2010, October 31). Green Living Answers. Retrieved October 31, 2010, from http://www.greenlivinganswers.com/wind-power/pros-and-cons-of-an-offshore-wind-farm

Q & A: Offshore wind farms. (2007, December 11). BBC News. Retrieved October 30, 2010 from http://news.bbc.co.uk/2/hi/uk_news/7136178.stm

Rhetoric vs. reality: Wind energy and birds. (2011). American Wind Energy Association. Retrieved May 20, 2011, from http://www.awea.org/newsroom/realstories/release_030111_2.cfm

Saber, K. (2009, February 12). "New ocean energy and offshore wind regulations announced." Alt Dot Energy. Retrieved August 8, 2010, from http://www.altdotenergy.com/2009/02/new-ocean-energy-and-offshore-wind-regulations-announced/

Seelye, Katharine Q. (2010, April 28). "Regulators approve first offshore wind farm in US, *The New York Times*. Retrieved August 8, 2010, from http://www.nytimes.com/2010/04/29/us/29wind.html

Shipman, K. (2011, January 4). Wind turbines may help crops, but key questions remain. FarmweekNow.com. Retrieved May 27, 2011 from http://www.farmweeknow.com/story.aspx/wind/turbines/may/help/crops/but/key/questions/remain/1/4520

Small Wind, Big Breakthrough. (2010, June 16). *Altenergymag.com*. Retrieved October 19, 2010, from http://www.altenergymag.com/news_detail.php?pr_id=15934

Small Wind Certification Council. (2010). About the Small Wind Certification Council. Retrieved October 18, 2010 from http://www.smallwindcertification.org

Thompson, A. (2007, February 14). Offshore Wind Farm Could Blow Away Energy Needs. *Live Science*. Retrieved October 29, 2010, from http://www.livescience.com/environment/070214_wind_farm.html

Tracton, S. (2009, July 15). Can wind farms change the weather? *The Washington Post*. Retrieved November 24, 2010, from http://voices.washingtonpost.com/capitalweathergang/2009/07/can_wind_farms_change_the_weat.html

Urban Wind Turbines. (Unknown date). *Urban Wind Energy*. Retrieved November 12, 2010, from http://www.urbanwindenergy.org.uk/index.asp?PageID=42

Waltz, Emily (2008, October 20). "Offshore wind may power the future," *Scientific American*. Retrieved August 8, 2010, from http://www.scientificamerican.com/article.cfm?id=offshore-wind-may-power-the-future

Watson, D. (2010). Wind turbines and the energy in wind. FT Exploring Science and Technology. Retrieved April 17, 2011, from http://www.ftexploring.com/energy/wind-enrgy.html

What is Betz Law? (2011). *Innovateus*. Retrieved April 17, 2011, from http://www.innovateus.net/earth-matters/what-betz-law

Wind data monitoring equipment on Texas state-owned lands. (2006). State Energy Conservation Office. Retrieved September 24, 2010, from http://www.seco.cpa.state.tx.us/re_wind_maps.htm

Wind energy FAQ. (2009). American Wind Energy Association. Retrieved August 30, 2010 from http://www.awea.org/faq/vawt.html

Wind energy resource atlas of the United States: Chapter 2. (n.d.). Renewable Resource Data Center. Retrieved on September 24, 2010, from http://rredc.nrel.gov/wind/pubs/atlas/chp2.html

Wind farms impacting weather: Environmental engineers detect turbines' turbulence. (2005, October 1). Science Daily. Retrieved November 24, 2010, from http://www.sciencedaily.com/videos/2005/1012-wind_farms_impacting_weather.htm

Wind Turbine Economics 2009: Everything You Need to Know. (2009, December 3). Renewable Energy Sources. Retrieved November 11, 2010, from http://www.renewable-energy-sources.com/2009/12/03/wind-turbine-economics-2009-everything-you-need-to-know/

Wind Turbine Tip Speed Ratio. (2010, September 12). REUK. Retrieved September 12, 2010 from http://www.reuk.co.uk/Wind-Turbine-Tip-Speed-Ratio.htm

Wind Web Tutorial: Offshore Wind. (2009). Wind Web Tutorial: Offshore Wind. AWEA. Retrieved June 29, 2011, from http://archive.awea.org/faq/wwt_offshore.html

Woofenden, I. (2009). *Wind power for dummies*. Indianapolis: Wiley Publishing, Inc.

Index

Figures and tables are indicated by f and t following page numbers.

D

E

T

X

Y